Generalizability Theory

MEASUREMENT METHODS FOR THE SOCIAL SCIENCES SERIES

Measurement Methods for the Social Sciences is a series designed to provide professionals and students in the social sciences and education with succinct and illuminating texts on measurement methodology. Beginning with the foundations of measurement theory and encompassing applications on the cutting edge of social science measurement, each volume is expository, limited in its mathematical demands, and designed for self-study as well as formal instruction. Volumes are richly illustrated; each includes exercises with solutions enabling readers to validate their understanding and confirm their learning.

Titles in this series . . .

1. GENERALIZABILITY THEORY: A PRIMER
   Richard J. Shavelson and Noreen M. Webb
2. FUNDAMENTALS OF ITEM RESPONSE THEORY
   Ronald K. Hambleton, H. Swaminathan, and H. Jane Rogers

Generalizability Theory

A PRIMER

Richard J. Shavelson
Noreen M. Webb

SAGE PUBLICATIONS
The International Professional Publishers
Newbury Park London New Delhi

For information address:

SAGE Publications, Inc.
2455 Teller Road
Newbury Park, California 91320

SAGE Publications Ltd.
6 Bonhill Street
London EC2A 4PU
United Kingdom

SAGE Publications India Pvt. Ltd.
M-32 Market
Greater Kailash I
New Delhi 110 048 India

Printed in the United States of America

Library of Congress Cataloging-in-Publication Data

Shavelson, Richard J.
Generalizability theory : a primer / Richard J. Shavelson, Noreen M. Webb.
p. cm. — (Measurement methods for the social sciences ; v. 1)
Includes bibliographical references and index.
ISBN 0-8039-3744-X (cloth). — ISBN 0-8039-3745-8 (pbk.)
1. Psychological tests—Evaluation. I. Webb, Noreen M. II. Title. III. Series.
BF176.S48 1991
150′.28′7—dc20 91-22006

92 93 94 15 14 13 12 11 10 9 8 7 6 5 4 3 2

Sage Production Editor: Diane S. Foster

We dedicate this book
to our mentor, colleague, and friend,
Lee J. Cronbach.

Contents

Series Editor's Foreword

In 1972, Lee Cronbach and his associates substantially broadened our conception of measurement reliability by introducing the theory of generalizability. Their classic volume effectively demonstrated that it was no longer necessary to restrict decomposition of variation in individuals' observed test scores to two components—variation attributed to true differences among individuals, and variation attributable to a conglomerate of systematic and random sources that included omitted variables, interactions between the elements of measurement and the persons measured, and ephemeral contributions to individual performance differences that were beyond measurement interest. Indeed, this latter component of variation could be dissected further to gain understanding of the systematic sources of variation that contributed to what we heretofore considered an undifferentiable mass, simply labeled "error." The generalizability theory enabled estimation of the relative magnitudes of various components of error variation and gain of a prescription for controlling error variation. The theory offered a powerful approach to assessing measurement consistency and offered the possibility of improving the reliability with which measurements were obtained while indicating the most efficient strategy for achieving desired measurement precision.

It has been 19 years since *The Dependability of Behavioral Measurements* (Cronbach, Gleser, Nanda, & Rajaratnam, 1972) was published, and a tailored computer program for calculating desired estimates of generalizability statistics (Crick & Brennan, 1982) was published 9 years ago. Generalizability methods, however, are far from pervasive in applied social science measurement. Indeed, articles in social science research journals contain traditional estimates of measurement reliability far more often than analyses of generalizability (assuming the researchers concerned themselves with consistency of measurement at all). One can only assume that, despite their power and promise, generalizability theory and methods are sufficiently complex that they have not yet entered the lexicon of techniques available to applied researchers in the social sciences.

In this volume, Shavelson and Webb have made the theory and methods of generalizability available to anyone who has mastered the most basic rudiments of the analysis of variance. With lucid prose, detailed illustrations and examples, and intuitively appealing descriptions of the logic underlying major concepts of generalizability theory, Shavelson and Webb have made this powerful set of techniques universally understandable and applicable. Thousands of social science researchers will no longer be forced to rely on outmoded reliability estimation procedures when investigating the consistency of their measurements. This *Primer* on generalizability provides a fitting beginning for the **Measurement Methods for the Social Sciences** series. It provides a clear introduction to an important measurement topic for social science researchers, a useful reference for professionals in social science measurement, and an important text for students of social science measurement. The authors have achieved the goals of the series with grace and acumen.

RICHARD M. JAEGER
University of North Carolina at Greensboro

Preface

A primer, according to the dictionary, is a "small introductory book on a subject." We intend this book to be just that: a small introductory book on *generalizability theory,* a psychometric theory of the dependability of behavioral measurements. Our goal is to provide the understanding and tools to permit the reader to use generalizability theory in everyday, practical situations. To this end, we develop constructs intuitively and follow this development with only the most necessary mathematical and statistical formalisms, lay the groundwork for more advanced developments, and justify practical applications. In developing the theory intuitively, we rely on analogies and on schematic and visual representatives, as well as clear verbal exposition. We map the technical material onto the intuitive development, to the extent possible. We provide diverse practical examples of the application of the theory, drawn largely from our own studies.

We make three important assumptions in writing the book. The first is that the readers are researchers and applied testing professionals with limited psychometric knowledge and skills. Their interest in generalizability theory is in its application to research and/or testing, not necessarily in its formal development, which is already available (Brennan, 1983; Cardinet & Tourneur, 1985; Cronbach et al., 1972).

The second assumption is that the readers are familiar with classical reliability theory. We assume that they understand and can apply such concepts as true score, error, reliability, test-retest reliability, parallel forms reliability, and internal consistency reliability.

The third assumption is that the readers are familiar with the analysis of variance (ANOVA). More specifically, we assume that the readers have a rudimentary understanding of how the ANOVA partitions variation in a set of data collected in a factorial design. Since this knowledge may be limited largely to the so-called fixed-effects ANOVA, random-effects and mixed-models ANOVAs are developed specifically in this book. We do not deal with hypothesis testing under random and mixed models.

RICHARD J. SHAVELSON
NOREEN M. WEBB

Acknowledgments

This *Primer* was written (and rewritten!) with the help and advice of colleagues and students. Lee Cronbach, Bob Brennan, and Donna Sundre challenged us, caught potentially misleading development of concepts, and provided expert guidance to us on the project. Weichang Li checked our computations and helped us correct errors. Gail Baxter read and constructively critiqued draft after draft. Dick Jaeger provided valuable comments and editorial suggestions. Graduate students in a summer psychometrics seminar—Gail Baxter, Janet Brown, Xiaohong Gao, Sharon Lesar, and James Valadez—identified problems and ambiguities and helped fix them. Last, but not least, Pat Skehan masterfully revised computer file after file, producing an aesthetically appealing manuscript . . . no mean task, with the number of equations involved.

1

Concepts in Generalizability Theory

Generalizability (G) theory is a statistical theory about the dependability of behavioral measurements. Cronbach, Gleser, Nanda, & Rajaratnam (1972) sketched the notion of dependability as follows:

> The score [on a test or other measure] on which the decision is to be based is only one of many scores that might serve the same purpose. The decision maker is almost never interested in the response given to the particular stimulus objects or questions, to the particular tester, at the particular moment of testing. Some, at least, of these conditions of measurement could be altered without making the score any less acceptable to the decision maker. . . . The ideal datum on which to base the decision would be something like the person's mean score over all acceptable observations. (p. 15)

Dependability, then, refers to the accuracy of generalizing from a person's observed score on a test or other measure (e.g., behavior observation, opinion survey) to the average score that person would have received under all the possible conditions that the test user would be equally willing to accept. Implicit in this notion of dependability is the assumption that the person's knowledge, attitude, skill, or other measured attribute is in a steady state; that is, we assume that any differences among scores earned by an individual on different occasions of measurement are due to one or more sources of error, and not to systematic changes in the individual due to maturation or learning.

A single score obtained on one occasion on a particular form of a test with a single administrator, then, is not fully dependable; that is, it is unlikely to match that person's average score over all acceptable occasions, test forms, and administrators. A person's score usually would be different on other occasions, on other test forms, or with different administrators. Which are the most serious sources of inconsistency or error? Classical test theory can estimate separately only one source of error at a time (e.g., variation in scores across occasions can be assessed with test-retest reliability).

The strength of G theory is that multiple sources of error in a measurement can be estimated separately in a single analysis. Consequently, in a manner similar to the way the Spearman–Brown "prophecy formula" is used to forecast reliability as a function of test length in classical test theory, G theory enables the decision maker to determine how many occasions, test forms, *and* administrators are needed to obtain dependable scores. In the process, G theory provides a summary coefficient reflecting the level of dependability, a generalizability coefficient that is analogous to classical test theory's reliability coefficient.

Moreover, G theory allows the decision maker to investigate the dependability of scores for different kinds of interpretations. One interpretation (the only kind addressed in classical test theory) concerns the relative standing of individuals: "Charlie scored higher than 95% of his peers." But what if the decision maker wants to know Charlie's absolute level of performance regardless of how his peers performed? G theory provides information on the dependability of scores for this kind of interpretation as well.

In this chapter, we develop the central concepts of generalizability theory, using concrete examples. In later chapters, procedures are presented for analyzing measurement data.

Generalizability and Multifaceted Measurement Error

The concept of dependability applies to either simple or complex universes to which a decision maker is willing to generalize. Our presentation will be made concrete with applications to a variety of behavioral measurements (e.g., achievement tests, behavior observations, opinion surveys) under increasingly broad definitions of the universe. We begin here with the simplest case, in which the universe is defined by one major source of variation, called a *facet* in G theory. Then we move to universes with two, three, or even more facets.

One-Facet Universes

From the perspective of G theory, a measurement is a sample from a universe of *admissible* observations, observations that a decision maker is willing to treat as interchangeable for the purposes of making a decision. (The decision may be practical, such as the selection of the highest scoring students for an accelerated program, or may be the framing of a scientific conclusion, such as the impact of an educational program on science achievement.) A one-facet universe is defined by one source of measurement error, that is, by a single facet. If the decision maker intends to generalize from one set of test items to a much larger set of test items, ITEMS is a facet of the measurement; the item universe would be defined by all admissible items. If the decision maker is willing to generalize from one test form to a much larger set of test forms, FORMS is a facet; the universe would be defined by all admissible test forms (e.g., all forms developed over the past 15 years). If the decision maker is willing to generalize from performance on one occasion to performance on a much larger set of occasions, OCCASIONS is a facet; the occasions universe would be defined by all admissible occasions (e.g., every day within a 3-month period). Error is always present when a decision maker generalizes from a measurement (a sample) to behavior in the universe.

Consider a typical school achievement test that consists of multiple-choice questions (called "items"), usually with four or five response alternatives. We consider here a simple scoring rule: right (1) or wrong (0). Based on the sample of items in the test, generalizations are made to students' "achievement," a generalization presumably not bound by the sample of items on the test.

If all of the items in the universe are equal in difficulty and an individual's score is roughly the same from one item to the next, then the person's performance on any sample of items will generalize to all items. If the difficulty of the items varies, however, a person's score will depend on the particular sample of items on the test. Generalization from sample to universe is hazardous. Item variability, then, represents a potential source of error in generalization. Items constitute a *facet* of the achievement measurement. If it is the only facet being considered, the set of admissible "items" is a single-faceted universe. The decision maker, of course, must decide which items are admissible.

As a concrete example, consider a science achievement test for fifth graders. The test contains 40, four-alternative multiple-choice items, scored (0, 1). Table 1.1 lists scores on a random sample of eight items that call for recall of factual information, reasoning with science concepts, interpretation of data or graphs, generalization from data or experimental set-ups, or the like. We use the mean of item scores as a person's score on this eight-item test. In practical testing, the total rather than the mean is used ordinarily. Throughout this *Primer* we shall use the mean of a set of observations as the summary "observed score." This usage greatly simplifies presentation of G theory. Any result obtained for mean scores is converted easily into a statement about total scores, so nothing is lost by basing formulas on means.

Users of achievement test information—such as school administrators, parents, policymakers, or the general public—are probably indifferent to the particular questions on the science test. They would be quite willing to substitute another set of items covering similar science facts, inferences, and interpretations, or covering different instances of the same facts, and so forth; that is, the users of test information are more interested in a student's *general* science achievement than the student's score on any particular set of items. The generalized achievement is represented by the score that would have been earned on a wide variety of test items that could have been used. Because items vary in difficulty across all students or for particular students, students scoring high on one sample of items may not score high on another sample of items. Test items, a *facet* of the measurement, is a potential source of error in generalization.

The item facet, then, can be represented as the wide variety of items that test users want to call "science test items." These items comprise the *universe* of test items. If we think of the universe, we could have a table just like Table 1.1 with two changes. First, instead of having just eight columns, we would have a very large number of columns covering all of the admissible items. (For brevity, we can use the infinity symbol ∞ for the number of items in the universe.) Second, the person mean is now the average over the whole set of items. This mean is the *universe score,* as contrasted with the "test score" in the last column of Table 1.1.

Ideally, test users want to know each person's universe score. Because this ideal datum is unknown, we want to know how accurate the generalization is from the particular set of items on the science test (Table 1.1) to all admissible items as an indicator of a student's science achievement.

A one-facet design has four sources of variability. One source of variability arises from systematic differences among students' achieve-

TABLE 1.1 Item Scores on Eight Items from the CTBS Science Achievement Test

	Item								
Person	*1*	*2*	*3*	*4*	*5*	*6*	*7*	*8*	*Person Mean*
1	0	1	0	0	0	1	0	1	0.375
2	1	0	1	0	0	0	0	1	0.375
3	1	1	1	0	0	0	0	0	0.375
4	1	1	0	0	1	0	0	1	0.500
5	1	1	1	1	0	0	0	1	0.625
6	1	1	1	1	1	1	1	1	1.000
7	1	0	1	0	0	1	1	0	0.500
8	1	0	1	0	0	0	1	1	0.500
9	1	0	0	0	0	1	1	1	0.500
10	0	1	0	0	1	1	0	0	0.375
11	0	0	0	1	1	1	0	0	0.375
12	0	0	1	0	0	0	0	0	0.125
13	1	1	1	1	1	1	1	1	1.000
14	0	0	0	0	0	1	1	1	0.375
15	0	0	1	0	0	0	0	1	0.250
16	1	1	1	0	0	1	0	0	0.500
17	0	1	0	0	0	0	0	0	0.125
18	1	0	0	0	0	1	1	1	0.500
19	0	0	0	0	0	1	1	0	0.250
20	0	1	0	0	0	0	0	0	0.125
Item Mean	0.55	0.50	0.50	0.20	0.25	0.55	0.40	0.55	0.4375

Source: The data are from *New Technologies for Assessing Science Achievement* by R. J. Shavelson, J. Pine, S. R. Goldman, G. P. Baxter, and M. S. Hine, 1989.

ment in science. We speak of this source of variability as the *object of the measurement*. Variability among the objects of measurement (usually persons, in social science measurement) reflects differences in their knowledge, skills, and so on.

The second source of variability arises from differences in the *difficulty* of the test items. Some items are "easy," some difficult, and some in between. To the extent that items vary in difficulty, generalization from the item sample to the item universe becomes less accurate.

The third source of variability arises from the educational and experiential histories that students bring to the test. For example, a test item on hamsters would be easier for a student who has raised them than for other students. The difference in ordering of students on different items constitutes, in analysis of variance terms, an interaction between persons and items. This match between a person's history and a particular item increases variability and increases the difficulty of generalizing from a student's score on the sample of eight science items to his or her average score over all possible items in the universe—the *universe score* (G theory's analog to classical theory's true score).

The fourth source of variability may arise out of randomness (e.g., a momentary lapse of a student's attention), other systematic but unidentified or unknown sources of variability (for example, different students take the test on different days), or both.

In sum, four sources of variability can be identified in the achievement test scores in Table 1.1: (a) differences among objects of measurement, (b) differences in item difficulty, (c) the person-by-item match, and (d) random or unidentified events. The third and fourth sources of variability, however, cannot be disentangled. With only one observation in each cell of the table, we do not know, after accounting for the first two sources, if differences between item scores reflect the person–item combination or "interaction" (as in analysis of variance), the random/unidentified sources of variability, or both. Consequently, we lump the third and fourth sources of variability together as a residual: the person-by-item interaction ($p \times i$) confounded with other unidentified sources we denote by the letter *e*.

The magnitude of the three types of variation can be estimated. G theory expresses the magnitude of variability in terms of variance components. In the one-facet design in Table 1.2 the variance components are σ^2_p, σ^2_i, and $\sigma^2_{pi,e}$. One purpose of G theory is to assess the major sources of variation so that unwanted variation can be reduced in collecting future data. In this application we are thinking about redesigning the measurement (e.g., lengthening the test) rather than evaluating the measures already made. For the one-facet design this means estimating and interpreting the magnitude of the variance components for items and the residual, compared to that for universe-scores. Chapters 2 and 3 present the statistical apparatus for estimating the magnitude of variation arising from differences in item difficulty and from the residual.

Two-Facet Universes

Social science measures are complex and, therefore, contain more than one facet. For example, the science test or a similar one might be

TABLE 1.2 Sources of Variability in a One-Facet Measurement (Science Achievement Example)

Source of Variability	*Type of Variability*	*Variance Notation*
Persons (p)	Universe score	σ_p^2
Items (i)	Conditions	σ_i^2
$p \times i$ interaction/ Unidentified or random	Residual	$\sigma_{pi,e}^2$

administered on a different date and the user might consider any of a number of dates equally suitable. The universe of admissible observations would be defined by the two facets—items and occasions—taken together; that is, the universe of admissible observations would be defined by all acceptable items that could be given at many points in time.

To make these abstractions concrete, we will use behavioral observations to exemplify a two-facet universe. The two facets will be observers and occasions. Observations of behavior are found commonly in basic research and in applied areas such as evaluations of job performances (e.g., performance of air traffic controllers or teachers), or appraisals in clinical settings (e.g., assessment centers, counseling transactions).

Typically, an observer watches persons and counts, rates, or scores their behavior. The frequency count, score, or rating given by a particular observer is used as an indication of the average frequency count, score, or rating that would have been obtained with many observers.

The greater the inconsistencies among observers in their ratings of behavior, the more hazardous the generalization from one observer's rating of behavior to the universe of interest. Where *items* constituted a facet of the achievement measurement described above, here *observers* is recognized as a facet.

In addition, repeated observations of behavior are made commonly in both research and applied settings, in recognition that behavior may vary from one occasion to the next. To the extent that behavior is inconsistent across occasions, generalization from the sample of behavior collected on one occasion to the universe of behavior across all occasions of interest is hazardous. Observation *occasions,* then, is a second facet.

TABLE 1.3 Data from Behavior Observation Measurement

	Occasion:	*1*		*2*	
Child	*Rater:*	*1*	*2*	*1*	*2*
1		0	1	1	2
2		3	4	1	2
3		2	2	1	0
4		1	2	0	1
5		1	2	2	1
6		4	4	3	4
7		1	1	2	1
8		2	2	0	0
9		1	1	1	2
10		1	1	1	0
11		1	2	1	1
12		1	2	1	1
13		2	1	1	1

Source: Data are from *Interaction Processes and Learning Among Third-Grade Black and Mexican-American Students in Cooperative Small Groups* by C. M. Kenderski, 1983.

A dissertation by Kenderski (1983) provides an example of the data generated by behavior observation. Nine-year-old children were observed while solving mathematics problems in class. The children's conversations while doing the work were tape-recorded. Raters read transcripts of the tapes and counted the number of times each child asked for help from other children. All children were observed (taped) on the same two occasions 3 weeks apart. The same two raters coded all transcripts. Table 1.3 provides the data from Kenderski's study.

Behavioral measurements with two facets, judging from Table 1.3, contain a number of different sources of variability. One source of variability, attributable to the object of measurement, is individual differences among children in their requests for help. This source of variability is considered universe-score variability.

A two-facet design has six other sources of variability, associated with the measurement *facets,* that create inaccuracies in generalizing from the particular sample of behavior in the measurement to the universe of admissible observations on the same child (see Table 1.4). Inconsistencies among raters will create problems in generalizing from the average frequency count provided by a sample of two raters to the average frequency counts provided by the entire universe of admissible raters. Conclusions about a child's help-seeking would depend on the

TABLE 1.4 Sources of Variability in the Two-Facet Observational Measurement

Source of Variability	*Type of Variation*	*Variance Notation*
Persons (p)	Universe-score variance (object of measurement)	σ_p^2
Raters (r)	Constant effect for all persons due to stringency of raters	σ_r^2
Occasions (o)	Constant effect for all persons due to their behavioral inconsistencies from one occasion to another	σ_o^2
$p \times r$	Inconsistencies of raters' evaluation of particular persons' behavior	σ_{pr}^2
$p \times o$	Inconsistencies from one occasion to another in particular persons' behavior	σ_{po}^2
$r \times o$	Constant effect for all persons due to differences in raters' stringency from one occasion to another	σ_{ro}^2
$p \times r \times o, e$	Residual consisting of the unique combination of p, r, o; unmeasured facets that affect the measurement; and/or random events	$\sigma_{pro,e}^2$

luck of the draw—a "liberal" rater rather than a "stringent" one. Note that stringency of a rater applies to *all* students or "subjects" in the population. Hence, we speak of this "main effect" (in analysis of variance parlance) as a constant effect, constant for all children.

Likewise, inconsistencies in the frequency of students' help-seeking from one occasion to the next make generalization from sample to universe difficult. Something that transpires on a particular occasion that affects all students in the same way may increase or decrease their help-seeking behavior. For example, at the end of the week when weekend events override other more mundane academic concerns, students may pay less attention to the task and be less likely to seek help. This situation would have a constant effect on all subjects participating in the study.

In addition, inconsistencies in raters' counts of help-seeking behavior may arise for particular subjects. For example, Rater 1 might be particularly liberal when rating subjects 4, 7, and 12—more liberal than other raters would be—while Rater 2 treats all subjects alike. Because of this variation, we speak of a person by rater interaction because only some people and some raters in combination produce a unique result.

Similarly, some people (but not all people) may need a lot of help on one occasion but not on another. This inconsistency does not apply to all persons but is "localized" with particular people. Hence, we speak of a person-by-occasion interaction.

The unique combination of rater leniency and occasion is also a source of variability. On one occasion, Rater 1 is liberal in counting behavior for all persons, while on another occasion he or she is not. Hence, the unique rater-by-occasion combination contributes to the variation of scores received by all persons and is called the rater-by-occasion interaction.

Finally, the last source of variability is the residual that includes the unique combination of person, rater, and occasion (the person-by-rater-by-occasion interaction), unmeasured sources of variation in the particular data collection design, and random events. As in the one-facet design, we use the symbol e to denote unmeasured and random sources of variation.

As with the one-facet design, a variance component (σ^2) describes how important each source is. Chapters 2 and 3 show how to estimate and interpret the variance components for all sources of variability.

Universes with Three or More Facets

The complexity of social science measures cannot always be captured by two facets. For example, the user of achievement test scores may want to generalize over three (or more) facets, such as items, occasions, and test administrators. We have seen already how student performance might vary over items and occasions. Test performance might vary also according to the clarity of an administrator's instructions or the tone the administrator sets for the test. Administrators, then, would constitute a third facet; generalizing from performance under one administrator to the average over all possible administrators may lead to error. The universe of admissible observations now would be defined by the three facets—items, occasions, and administrators—taken together; that is, the universe of admissible observations would be defined by all possible items that could be given by all possible administrators at all possible points in time. The broader the universe of admissible observations, the greater the possibility of making an error in generalizing from sample to universe.

Designs with Crossed and Nested Facets

In the one-facet science test example presented earlier, all persons were administered all items. We speak of a measurement in which all conditions of one facet (e.g., items) are observed with all conditions of another source of variation (persons) as a *crossed* measurement. In this design, then, persons were crossed with items. Similarly, in the two-facet behavior observation measure described earlier, each rater counted the number of times each child asked questions on each occasion. In this design, the three sources of variability—persons, raters, and occasions—were all crossed.

In the child behavior study, raters coded transcripts of tapes of classwork, so it was convenient to have all raters code the transcripts from multiple occasions. Suppose instead that raters observed the children's behavior in the classroom itself rather than working from tapes. For the design to be crossed, all raters would have to visit the classroom on all occasions. It may be more convenient, however, for different raters to be present on different occasions. In this case, the rater facet would *not* be crossed with the occasion facet. Rather, we speak of the rater facet as *nested* within the occasion facet. One facet is said to be nested within another facet when two or more conditions of the nested facet (raters) appear with one and only one condition of another facet (occasions). Two or more raters are present on one occasion, a different set of raters are present on another occasion, and so on. Generalizability theory can be applied to fully crossed designs or to designs with a mixture of crossed and nested facets.

Random and Fixed Facets

Samples are considered "random" when the size of the sample is much smaller than the size of the universe, and the sample either is drawn randomly or is considered to be exchangeable with any other sample of the same size drawn from the universe. To decide whether a sample of conditions of a facet might be treated as random and, consequently, whether the facet could be treated as random, the question should be asked, "Am I willing to *exchange* the conditions in the sample for any other same-size set of conditions from the universe?" If the answer is yes, the facet can be classified as random. For example, if decision makers were willing to exchange the 40 items on the Science Achievement Test for another sample of 40 items, the facet might be treated reasonably as random (for additional details, see Shavelson & Webb, 1981).

In other cases, however, the conditions of the facet exhaust the conditions in the universe to which the researchers want to generalize. For example, achievement tests often have multiple subtests covering such content as mathematics, science, social studies, language arts, and so on. Here, exchangeability is not the issue; the number of conditions of the subtest facet equals the number of conditions in the universe of generalization. Consequently, we treat the subtest facet as a *fixed* facet (analogous to a fixed factor in ANOVA).

G theory distinguishes between measurement facets as random or fixed. As will be shown in later chapters, whether a facet is treated as random or fixed will have consequences for the generalizability of the measurement.

Generalizability and Decision Studies

In G theory a distinction is made between *generalizability* (G) studies and *decision* (D) studies. The purpose of a G study is to anticipate the multiple uses of a measurement and to provide as much information as possible about the sources of variation in the measurement. (In this chapter, to this point, then, we have been describing G studies.) A G study should attempt to identify and to incorporate into its design as many potential sources of variation as possible. Or, put another way, the G study should define the *universe of admissible observations* as broadly as possible.

A D study makes use of the information provided by the G study to design the best possible application of the social science measurement for a particular purpose. In planning a D study, the decision maker does the following: (a) defines a *universe of generalization*—the number and breadth of facets that he or she is willing to generalize over; (b) specifies the proposed interpretation of the measurement—relative or absolute (see next section); the proposed interpretation defines measurement error and thereby identifies the sources of error of greatest concern; and (c) uses the information from the G study about the magnitude of the various sources of measurement error to evaluate the effectiveness of alternative designs for minimizing error and maximizing reliability. This evaluation is done in a manner analogous to the Spearman-Brown prophecy formula in classical test theory. By increasing the number of conditions of a facet in a measurement (e.g., increasing the number of observers who rate each behavior), the error contributed by that facet can be decreased, much as adding items to a test decreases error (and increases reliability) in classical test theory. In this way, a cost-efficient design for the application of the social science measure can be developed.

Relative and Absolute Decisions

In G theory, how generalizable a measure is depends on how the data will be used in the decision study. Measurements in the social sciences are characteristically used in two ways: (a) to rank order individuals (or groups); and (b) to index an individual's (or a group's) absolute level of knowledge, skill, attitude, or strength of opinion. For example, the Science Test can be used to order students from low to high scorers. Students' scores can be interpreted in terms of how far above or below others they fall. For example, "Charlie scored higher than two thirds of his peers on the test." Moreover, rank ordering of individuals is the focus of studies investigating the relationship between two variables, such as verbal ability and science achievement. In these cases, attention is paid to the standing of individuals relative to one another. Such use of measurements contributes to *relative decisions.*

In contrast, measurements are used also to index the absolute level of an individual's performance without regard to how well or poorly his or her peers performed. For example, tests of typing speed measure words per minute, and the score a person receives does not depend on how well his or her peers performed. Likewise, a written drivers' examination often has a passing score, a standard, set at, say, 80% of the items answered correctly. Passing or failing the exam depends on the number of items a person answered correctly, not on how other people performed on the test that day. Finally, a knowledge domain can be specified, such as single-digit subtraction, and items sampled from the domain. A person's score on the test provides an estimate of the percentage of all such items that the person could perform correctly in the domain, independent of how his or her peers performed. In each of these cases, attention focuses on the absolute level of an individual's performance, not on relative standing. Decisions based on the absolute level of performance are called *absolute decisions.* As will be discussed in the next chapters, the kind of decision to be made has a bearing on the definition of measurement error and on the magnitude of error variance.

Generalizability Coefficient

Although G theory focuses on the sources of variation in social science measurements that contribute to error, it also provides a reliability coefficient called, not surprisingly, a *generalizability (G) coefficient.* The G coefficient shows how accurate the generalization is from a person's *observed score,* based on a sample of the person's behavior, to his or her universe score. Like classical test theory's reliability coefficient, the generalizability coefficient reflects the proportion of variability in individuals' scores that is systematic; that is, attributable to universe-score (cf. true-score) variability. Applied to the science test example, the G coefficient reflects the proportion of observed test-score variance attributable to systematic differences in students' science knowledge, that is, their universe-score variability.

The definition of the G coefficient depends on how a measurement is to be used. Because error variance is different for relative and absolute decisions, the magnitude of the G coefficient will depend on the kind of decision to be made. The details will be presented in later chapters.

Summary

A test score is but a single sample from an indefinitely large universe of scores that a decision maker might be equally willing to use as an index of a person's ability, attitude, performance, and so on. A question, therefore, arises as to the accuracy of the generalization from a small sample to the large universe. "Reliability" resolves into the accuracy of an inference from an individual's test score to the person's average score over the universe ("universe score"). The universe might be fairly simple, involving (say) alternative forms of a test; generalization would be from a person's score on one form to the person's average score across all possible forms. Or the universe might be complex, involving forms, occasions, administrators, and the like. G theory enables the analyst to isolate different sources of variation in the measurement (e.g., forms or occasions) and to estimate their magnitude using the analysis of variance. The variance components estimated in the G study can be used then to design time- and cost-efficient applications of the measurement for decision-making purposes (the "decision [D] study").

Exercises

1. Define dependability of measurement.
2. Some flight training schools now have computer simulators in which pilot trainees can practice "flying" a plane. In a test of "landings," 25 flight trainees were scored on the quality of their landings in 10 simulated flights.
   a. List the sources of variation. Identify the object of measurement and facet(s).
   b. What is the design of this study?
   c. List the variance components associated with this design.
3. Four supervisors from public schools for hearing-impaired students rated 30 teachers of the hearing impaired on 40 competencies considered most critical to the effective teaching of hearing-impaired students.
   a. List the sources of variation. Identify the object of measurement and facet(s).
   b. What is the design of this study?
   c. List the variance components associated with this design.
4. How does an absolute decision differ from a relative decision?
5. In the following examples, identify whether the behavioral measurement interpretations are based on relative or absolute decisions.
   a. To pass the California Bar Exam, 70% of the items need to be answered correctly.
   b. Prescott Farm Insurance Company administers a word-processing test to 25 applicants. The five top-scoring applicants will be selected for the job positions.
   c. A 10-item test is used to decide whether a student has mastered a unit in a mathematics curriculum.
   d. Five judges rate 10 women on a 6-point scale for gymnastic abilities to be considered for the winter Olympics. The three women with the highest ratings will qualify to be on the competing U.S. team.
6. What is the relationship between a G study and a D study?
7. How should researchers doing G studies and D studies define their respective universes?

Answers to Exercises

1. The dependability of measurement refers to the trustworthiness of generalization from a particular set of items (occasions, raters, etc.) to all admissible items.
2. a. Flight trainees (object of measurement), landing trials (facet).
   b. In this design, flight trainees (p) are crossed with landing trials (t). We denote this design as $p \times t$.
   c. Three variance components can be estimated in this design: flight trainees (p), landing trials (t), and the interaction between them, confounded with unmeasured variation (pt,e).
3. a. Teachers (object of measurement), supervisor (facet), and competencies (facet).
   b. In this design, teachers (p), supervisors (s), and competencies (c) are all crossed. We denote this design as $p \times s \times c$.
   c. σ^2_p, σ^2_s, σ^2_c, σ^2_{ps}, σ^2_{pc}, σ^2_{sc}, $\sigma^2_{psc,e}$
4. An absolute decision concerns individuals' level of knowledge, skill, attitude, and so on, without regard to the performance of others. A relative decision concerns the relative standing of individuals in a population.
5. a. *Absolute decision.* For absolute decisions, the level of their performance is taken into account.
   b. *Relative decision.* For relative decisions, only those sources of error affecting the relative standing of individuals contribute to measurement error.
   c. *Absolute decision.* For absolute decisions, the level of their performance is taken into account.
   d. *Relative decision.* For relative decisions, only those sources of error affecting the relative standing of individuals contribute to measurement error.
6. A D study uses the information provided by the G study to design the best possible measurement for a particular purpose.
7. G studies should define the universe of admissible observations as broadly as possible in order to assess the magnitudes of sources of error as broadly as possible. D studies define the universe of generalization as the breadth of facets over which the researcher will generalize while aiming for a cost-efficient design for an applied measure. Consequently, a single G study can supply the information needed by several D studies.

2

Statistical Model Underlying Generalizability Theory

Generalizability theory is to measurement what the ANOVA is to substantive research. Just as the researcher attempts to identify and estimate the effects of potentially important independent variables, the measurement specialist using G theory attempts to identify and estimate the magnitude of the potentially important sources of variation in a social science measurement—variation due to universe scores and multiple sources of error.

More specifically, just as ANOVA partitions an individual's score into the effects for the independent variables, their combinations (interactions), and error, G theory uses the factorial ANOVA to partition an individual's score into an effect for the universe-score (for the object of measurement), an effect for each facet or source of error, and an effect for each of their combinations.

In this chapter we present a statistical model for partitioning the variability among individuals' test (and other) scores.[1] The model is a familiar one. It is none other than the analysis of variance (ANOVA).

The ANOVA accomplishes this partitioning by working with variances. Consequently, in substantive research, the total variation in scores is partitioned into independent sources of variability due to each independent variable, their interactions, and error. For example, in a two-factor design with independent variables A and B, the ANOVA partitions variability among scores into an effect for A, an effect for B, their interaction ($A \times B$), and error (within-cell variability).

Analogously, the ANOVA can be used to partition variability among persons' achievement item scores into an effect for persons (universe-score variance in G theory), an effect for items (variability due to item difficulty), and a residual that includes the person-by-item interaction,

TABLE 2.1 Sources of Variability in a Single-Facet Measurement (Science Achievement Example) [a]

Source of Variability	*Type of Variability*	*Variance Notation*
Persons (p)	Universe score	σ_p^2
Items (i)	Error	σ_i^2
Residual (pi,e)	Error	$\sigma_{pi,e}^2$

a. Modification of Table 1.2 reproduced from chapter 1.

other systematic sources of error not explicitly identified in the persons-by-items design, and random variation (Table 2.1).

We denote the *observed score* for any person (p) on any item (i) as X_{pi}. Each cell of the persons-by-items matrix, then, contains an X with subscripts denoting the particular person–item combination.

Any score X_{pi} can be expressed as a sum involving three parameters—μ_p, μ_i, and μ. The *universe score,* denoted as μ_p, is defined as a person's average score over the entire item universe. We take this average to characterize a person's science achievement, as opposed to a person's average over the items on a test, which consists only of a sample of items from the universe. The i subscript is dropped because μ_p is the average over all items for person p.

More formally, we define a person's universe score, μ_p, as the *expected value* (E) of the random variable, X_{pi}, across items:[2]

$$\mu_p \equiv \mathrm{E}_i\,(X_{pi}) \qquad [2.1]$$

The expected value (E) of a random variable such as X_{pi} is, simply put, the long-run average. The long-run average is the average over k (the number of items) as that number approaches infinity (i.e., the limit as $k \to \infty$):

$$\mathrm{E}_i X_{pi} = \lim_{k \to \infty} \frac{1}{k} \sum_{i=1}^{k} X_{pi} \qquad [2.2]$$

Just as Σ tells us to take the sum of the term that follows it, the E symbol instructs us to take the expected value (long-run average) of the term that follows it.

The second parameter, μ_i, is the population mean for item i. It is defined as the expected value of X_{pi} over persons:

$$\mu_i \equiv \mathrm{E}_p X_{pi} \qquad [2.3]$$

The third parameter, μ, is the mean over both the population and universe, that is, the "grand mean" over the population and universe:

$$\mu \equiv \mathrm{E}_p \, \mathrm{E}_i X_{pi} \qquad [2.4]$$

The parameters μ_p, μ_i, and μ are not observable. An individual's responses to all items in the universe are never available, nor are the responses of all persons in the population to a given item. We are now in a position to define an observed score, X_{pi}, in terms of these parameters. The next chapter will show how the parameters can be estimated from sample data, using the ANOVA.

An observed score for one person on one item (X_{pi}) can be decomposed as follows:

$$\begin{aligned} X_{pi} = \; & \mu && \text{[grand mean]} \qquad [2.5] \\ & + \mu_p - \mu && \text{[person effect]} \\ & + \mu_i - \mu && \text{[item effect]} \\ & + X_{pi} - \mu_p - \mu_i + \mu && \text{[residual]}^3 \end{aligned}$$

In words, a person's score in a one-facet, crossed design depends on four components:

1. The grand mean, a constant for all people, positions the score on the particular scale of measurement.
2. The person effect shows the distance between an individual's universe score (μ_p) and the grand mean. A positive person effect indicates that the person scored higher than average; a negative effect means that a person scored lower than average.
3. The item effect shows the difficulty of the particular item. A positive item effect indicates that the item is easier than average (i.e., more people answered it correctly than the average item); a negative item effect means

that the item is more difficult than average (fewer people answer it correctly than the average score across all items).

4. Finally, the residual reflects the influence of the $p \times i$ interaction, other systematic sources of error not expressly included in the one-facet measurement, and random events.

Variance Components

Each effect other than the grand mean in Equation 2.5 has a distribution. (The grand mean is a constant, and so its variance is zero.) Each distribution has a mean of zero, and its own variance σ^2, called the variance component. Consider first, the person effect. The mean of the person effects, over all persons, has to be zero:

$$E_p(\mu_p - \mu) = E_p(\mu_p) - E_p(\mu) = \mu - \mu = 0 \qquad [2.6]$$

The variance of the person effects is labeled σ_p^2 and is called the variance component for persons, or universe-score variance. In formal terms, it is

$$\sigma_p^2 = E_p(\mu_p - \mu)^2 \qquad [2.7]$$

Equation 2.7 says that σ_p^2 is equal to the average (over the population of persons) of the squared deviations of the persons' universe scores from the grand mean. The variance component for persons shows how much persons differ from one another in their achievement or, equivalently, how much they differ from the grand mean.

The mean and variance component for items can be defined similarly. The mean of the item effect is zero, and its variance is σ_i^2. The variance component for items reflects the variance of constant errors (see chapter 1) associated with the difficulty levels of items in the universe; that is, the variance component for items shows how much items differ from one another in difficulty.

The final effect, the residual, also has a mean of zero and a variance denoted as $\sigma_{pi,e}^2$. In the residual, the $p \times i$ interaction effect is confounded with unmeasured or unsystematic variability. With only one observation per cell of the $p \times i$ data matrix (see Table 1.1), the interaction cannot be disentangled from "within-cell" variability. The notation for the

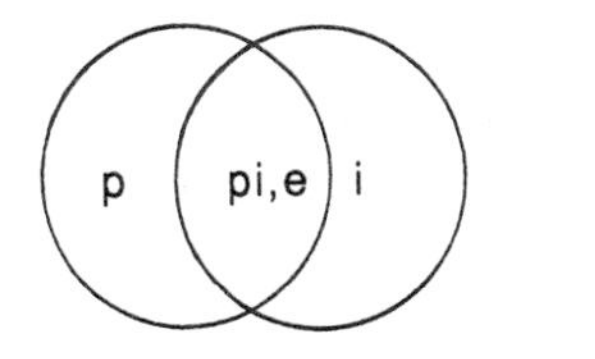

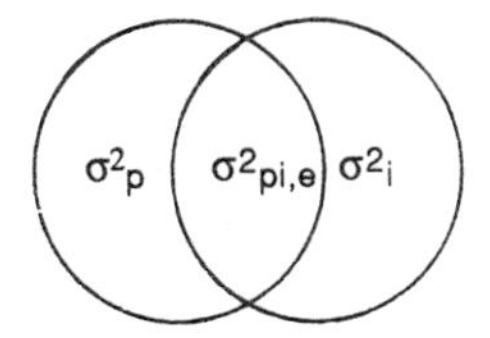

(a) Sources of variability (b) Variance components

Figure 2.1. Venn Diagrams for a One-Facet, Crossed $p \times i$ Design

residual variance component ($\sigma^2_{pi,e}$) reflects this confounding of the $p \times i$ interaction with these other sources of variation (here called e). The $p \times i$ effect reflects the fact that not all people find the same items easy or difficult. The e effect reflects, in part, unsystematic or random error sources. It may arise, for example, when a person breaks his or her pencil during the test and loses time on later items, causing his or her score to be lower than it should be. The e effect also includes systematic influences from facets not explicitly included or controlled in the one-facet G study. For example, individuals might take the examination in rooms with different noise levels. Persons in noisy rooms may have lower scores than they would have in quiet rooms. Although the testing condition is a systematic source of variation that could be included as a facet in a study, it is not a factor in a $p \times i$ design, and nothing is learned about its influence on test scores.

The variance of the collection of observed scores, X_{pi}, over all persons and items in the universe is the sum of the three variance components:

$$\sigma^2 (X_{pi}) = \sigma^2_p + \sigma^2_i + \sigma^2_{pi,e} \qquad [2.8]$$

In words, the variance of item scores can be partitioned into independent sources of variation due to differences between persons, items, and the residual.

A useful visual aid for decomposition of the total variance of observed scores into its constituent variance components is the Venn diagram. Figure 2.1, part (a), shows the Venn diagram for a crossed, one-facet design. The intersecting circles for p and i denote the crossing of persons and items. Figure 2.1, part (b), shows the analogous decomposition in terms of variance components. It is important to note that the magnitudes of the circles or areas within them do not represent the magnitudes of the variance components. Rather, Venn diagrams are useful for identifying sources of variance in various designs.

Multifaceted Designs

The partitioning of observed scores into their effects and the decomposition of the variance of observed scores into variance components for the separate effects easily extends to measurements with additional facets. Consider the two-facet, crossed behavior observation described in chapter 1. Children's help-seeking behavior while working on mathematics problems was coded by two raters on two different occasions. Children are the object of measurement; raters and occasions constitute sources of unwanted variation in the measurement. Raters and occasions are considered to be randomly sampled from indefinitely large universes; they constitute random facets. The behavior of all children was coded by both raters on both occasions so that the design is completely crossed. We denote this persons (p) by raters (r) by occasions (o) design as $p \times r \times o$.

X_{pro} represents the observed rating of one person (p) by one rater (r) on one occasion (o). (Although children and raters are both people, we use the term *persons* for the object of measurement whenever the focus of the measurement is on people.) Any person's observed score can be decomposed into components, as was done for a single-facet measurement in Equation 2.5. This design has eight components: (a) grand mean, (b) person effect, (c) rater effect, (d) occasion effect, (e) person-by-rater interaction effect, (f) person-by-occasion interaction effect, (g) rater-by-occasion interaction effect, and (h) residual (consisting of the person-by-rater-by-occasion interaction effect confounded with other unidentified or random influences). More formally, the observed score in a two-facet measurement can be decomposed as follows:

$$
\begin{aligned}
X_{pro} = \quad & \mu && \text{[grand mean]} \qquad [2.9] \\
+ \; & \mu_p - \mu && \text{[person effect]} \\
+ \; & \mu_r - \mu && \text{[rater effect]} \\
+ \; & \mu_o - \mu && \text{[occasion effect]} \\
+ \; & \mu_{pr} - \mu_p - \mu_r + \mu && \text{[person-by-rater effect]}
\end{aligned}
$$

$$+ \ \mu_{po} - \mu_p - \mu_o + \mu \qquad \text{[person-by-occasion effect]}$$
$$+ \ \mu_{ro} - \mu_r - \mu_o + \mu \qquad \text{[rater-by-occasion effect]}$$
$$+ \ X_{pro} - \mu_{pr} - \mu_{po} - \mu_{ro}$$
$$+ \ \mu_p + \mu_r + \mu_o - \mu \ \text{[residual effect]}$$

Each component except the grand mean has a distribution. For example, behavior will vary from one child to another, raters will differ in their coding of that behavior, and so on. Each distribution has a mean of zero, and a variance. The total variance over the universe and population for the two-facet, crossed measurement is equal to the sum of the variance components for the seven effects:

$$\sigma^2 (X_{pro}) = \sigma^2_p + \sigma^2_r + \sigma^2_o + \sigma^2_{pr} + \sigma^2_{po} + \sigma^2_{ro} + \sigma^2_{pro,e} \quad [2.10]$$

The Venn diagrams in Figure 2.2 show the seven sources of variance and the corresponding variance components.

The variance component for persons (σ^2_p), also called universe-score variance, shows how much children differ in their help-seeking behavior. The variance component for raters (σ^2_r) shows the extent to which some raters "see" more help-seeking behavior than others. The variance component for occasions (σ^2_o) shows the extent to which the average level of help-seeking behavior is higher on one occasion than another. The variance component for the person-by-rater interaction (σ^2_{pr}) shows the extent to which the relative standing of children in their help-seeking behavior changes from one rater to another. The variance component for the person-by-occasion interaction (σ^2_{po}) shows the inconsistency in relative standing of children from one occasion to the next. The variance component for the rater-by-occasion interaction (σ^2_{ro}) shows the inconsistency in raters' average ratings of students from one occasion to the next. Finally, the residual variance component ($\sigma^2_{pro,e}$) reflects the three-way interaction between persons, raters, and occasions confounded with unmeasured sources of variation. The unmeasured sources of variation may be unsystematic or random (e.g., a child feeling ill on one occasion and not asking for help). Or they may be systematic sources of variability that were not measured in this study (e.g., different children working on different mathematics topics).

The next chapter shows how to estimate and interpret variance components for one- and two-facet crossed designs. Later chapters will treat more complex designs.

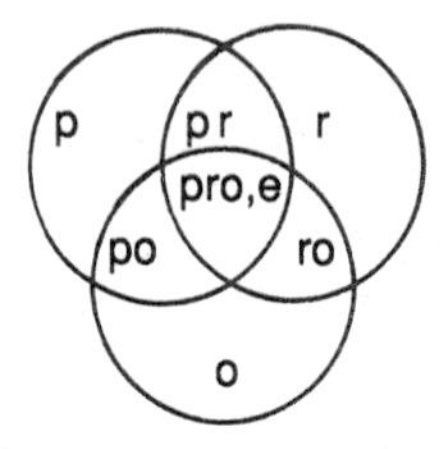

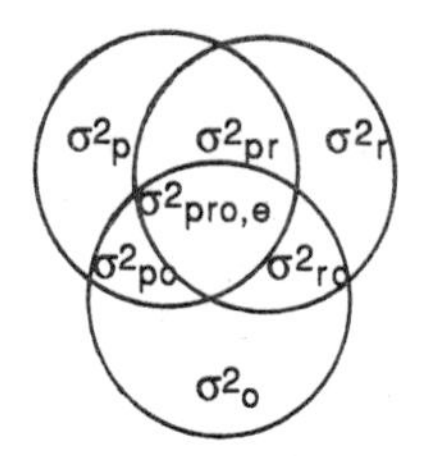

(a) Sources of variability (b) Variance components

Figure 2.2. Venn Diagrams for a Two-Facet, Crossed $p \times r \times o$ Design

Summary

Generalizabilty theory assumes that a person's observed score consists of a universe score (analogous to classical test theory's true score) and one or more sources of error. For example, an observed score for a person on a test item can be decomposed into four components: (a) a constant that reflects the scale of measurement—the grand mean in the universe, μ; (b) the effect for the person's universe score, $(\mu_p - \mu)$; (c) the effect for the particular item $(\mu_i - \mu)$; and (d) the effect of a residual that consists of the person-by-item interaction and other unidentified sources of variability $(X_{pi} - \mu_p - \mu_i + \mu)$.

G theory partitions variation across observed scores into separate sources corresponding to those in the observed-score model. For the $p \times i$ example, G theory partitions variation into three sources—persons, items, and the residual. More formally, G theory defines a variance component, σ^2, for each source of variation in observed scores—in this example, variance components for persons (σ_p^2), items (σ_i^2), and the residual $(\sigma_{pi,e}^2)$. By partitioning variability in this manner, G theory enables the analyst to pinpoint the major source(s) of measurement error, to estimate the total magnitude of error, and to form a reliability coefficient.

Exercises

1. In a G study of military job performance, 150 Marine Corps riflemen (p) were observed by two raters (r) as the riflemen performed 35 infantry tasks (t) sampled from the range of infantry job requirements. In this design all sources of variation are crossed: $p \times r \times t$. Give the formal decomposition of the observed score X_{prt}.
2. Give the formal definition of the variance of each effect in Exercise 1.
3. Interpret the variance of the rater effect in Exercise 2.
4. Identify and interpret the variance components associated with the $p \times r \times t$ design.
5. In the Marine Corps riflemen study, what factors contribute to the *prt,e* term?

Answers to Exercises

1.
$$
\begin{aligned}
X_{prt} = \;& \mu && \text{[grand mean]}\\
&+ \mu_p - \mu && \text{[person effect]}\\
&+ \mu_r - \mu && \text{[rater effect]}\\
&+ \mu_t - \mu && \text{[task effect]}\\
&+ X_{pr} - \mu_p - \mu_r + \mu && \text{[person-by-rater effect]}\\
&+ X_{pt} - \mu_p - \mu_t + \mu && \text{[person-by-task effect]}\\
&+ X_{rt} - \mu_r - \mu_t + \mu && \text{[rater-by-task effect]}\\
&+ X_{prt} - \mu_{pr} - \mu_{pt} + \mu_{rt} && \text{[residual effect]}\\
&\quad + \mu_p + \mu_r + \mu_t - \mu &&
\end{aligned}
$$

2.
$$
\begin{aligned}
\sigma^2_p &= E_p(\mu_p - \mu)^2\\
\sigma^2_r &= E_r(\mu_r - \mu)^2\\
\sigma^2_t &= E_t(\mu_t - \mu)^2\\
\sigma^2_{pr} &= E_p E_r(\mu_{pr} - \mu_p - \mu_r + \mu)^2\\
\sigma^2_{pt} &= E_p E_t(\mu_{pt} - \mu_p - \mu_t + \mu)^2\\
\sigma^2_{rt} &= E_r E_t(\mu_{rt} - \mu_r - \mu_t + \mu)^2\\
\sigma^2_{prt,e} &= E_p E_r E_t(X_{prt} - \mu_{pr} - \mu_{pt} - \mu_{rt} + \mu_p + \mu_r + \mu_t - \mu)^2
\end{aligned}
$$

3. The variance of the rater effect—$\sigma^2_r = E_r(\mu_r - \mu)^2$—is the long-run average (over all raters in the universe) of the squared deviations of the

rater means (μ_r) from the grand mean (μ). Each squared deviation—$(\mu_r - \mu)^2$—is the distance between one rater's mean score (over all persons and tasks) and the mean of all raters' means. The average of all raters' squared deviations, then, is the average squared distance between raters' means and the grand mean. It shows how much raters differ in their mean ratings, on the average.

4. σ_p^2 = universe-score variance. Shows how much Marines differ in their proficiency at infantry tasks.

   σ_r^2 = main effect for raters. Shows whether some raters are more lenient than others in their scoring of Marines.

   σ_t^2 = main effect for tasks. Shows whether some tasks are more difficult than others.

   σ_{pr}^2 = interaction between Marines and raters. Shows whether the relative standing of Marines differs across raters, averaging over tasks.

   σ_{pt}^2 = interaction between Marines and tasks. Shows whether the relative standing of Marines differs across tasks, averaging over raters.

   σ_{rt}^2 = interaction between raters and tasks. Shows the inconsistency of raters' average ratings of Marines from one task to the next.

   $\sigma_{prt,e}^2$ = three-way interaction plus remaining unmeasured error.

5. (1) The $p \times r \times t$ interaction.

   (2) Systematic variation due to sources not controlled for in this design. Example: Different Marines use different equipment to perform the tasks.

   (3) Unsystematic variation due to sources that cannot be controlled. Example: A piece of equipment jams unexpectedly, and Private Smith cannot complete the task in the time allotted.

Notes

1. In this chapter we assume a fully crossed, random-effects ANOVA. In later chapters we consider designs with fixed and nested facets.

2. The symbol ≡ is read "defined as."

3. This last term is a bit curious. It says that the residual is whatever is left over in the observed score after accounting for the grand mean on the construct, a person's universe score effect, and the effect of the difficulty of the particular item. Even more curious is the fact that μ is added onto the observed score! This is mathematical, not intuitive:

$$X_{pi} - \mu - (\mu_p - \mu) - (\mu_i - \mu) = X_{pi} - \mu - \mu_p + \mu - \mu_i + \mu = X_{pi} - \mu_p - \mu_i + \mu$$

3

Generalizability Studies with Crossed Facets

The previous chapter presented the formal model for partitioning scores into components. This chapter shows how applying analysis of variance to data from a crossed G study provides estimates of the variance components and goes on to show how the results are interpreted.

ANOVA Estimation and Interpretation of Variance Components: One-Facet G Studies

The G study for science achievement scores had a crossed $p \times i$ design. In the example, eight items were administered to 20 individuals; both items and persons are regarded as random samples from the universe of items and population of persons, respectively (Table 1.1). ANOVA can be used to estimate the variance components described in chapter 2.

ANOVA Estimation of Variance Components

Recall the familiar partitioning of the total sum of squares (SS_t) for a simple $p \times i$ design into sums of squares for persons, items, and the residual: $SS_t = SS_p + SS_i + SS_{pi,e}$. Dividing the sums of squares (SS) by their respective degrees of freedom (df) yields the mean squares (MS; see Table 3.1).

The last column moves beyond the familiar analysis to add important information: expected mean squares. The expected mean square (EMS) is the value of the mean square that would be obtained, on average, by

TABLE 3.1 ANOVA Table for a Persons-by-Items Design

Source of Variation	*Sums of Squares*	*df*	*Mean Squares*	*Expected Mean Squares*
Persons (p)	SS_p	$n_p - 1$	$MS_p = SS_p / df_p$	$E(MS_p) = \sigma^2_{pi,e} + n_i \sigma^2_p$
Items (i)	SS_i	$n_i - 1$	$MS_i = SS_i / df_i$	$E(MS_i) = \sigma^2_{pi,e} + n_p \sigma^2_i$
pi,e	$SS_{pi,e}$	$(n_p - 1)(n_i - 1)$	$MS_{pi,e} = SS_{pi,e} / df_{pi,e}$	$E(MS_{pi,e}) = \sigma^2_{pi,e}$

repeatedly analyzing samples from the same population and universe with the same design. The expected mean squares provide weighted sums of variance components as follows (for derivations of EMS equations, see Kirk, 1982):

$$E(MS_p) = \sigma^2_{pi,e} + n_i \sigma^2_p \quad [3.1]$$

$$E(MS_i) = \sigma^2_{pi,e} + n_p \sigma^2_i \quad [3.2]$$

$$E(MS_{pi,e}) = \sigma^2_{pi,e} \quad [3.3]$$

These equations are solved to obtain estimates of each variance component. We replace the expected mean squares with the corresponding observed mean squares and substitute the symbol $\hat{\sigma}^2$ for each σ^2:

$$MS_p = \hat{\sigma}^2_{pi,e} + n_i \hat{\sigma}^2_p \quad [3.4]$$

$$MS_i = \hat{\sigma}^2_{pi,e} + n_p \hat{\sigma}^2_i \quad [3.5]$$

$$MS_{pi,e} = \hat{\sigma}^2_{pi,e} \quad [3.6]$$

The "hat" (^) is a reminder that we are obtaining a sample estimate, not the exact value of σ^2.

These equations are solved using simple algebra. It is easiest to work from the "bottom up"; that is, solving for the residual first and working backward to obtain $\hat{\sigma}^2_i$ and $\hat{\sigma}^2_p$.

TABLE 3.2 ANOVA Estimates of Variance Components for the Science Achievement Data

Source of Variation	*Sums of Squares*	*df*	*Mean Squares*	*Estimated Variance Components*	*Percentage of Total Variance*
Persons (p)	8.6250	19	0.4539	0.0305	12
Items (i)	2.7750	7	0.3964	0.0093	4
pi,e	27.9750	133	0.2103	0.2103	84

The first step (Equation 3.6) is obvious. The mean square for the residual is the sample estimate of the corresponding variance component:

$$\hat{\sigma}^2_{pi,e} = \text{MS}_{pi,e} \quad [3.7]$$

The equation for items (Equation 3.5) can be rewritten this way:

$$\hat{\sigma}^2_i = (\text{MS}_i - \hat{\sigma}^2_{pi,e}) / n_p \quad [3.8]$$

The ANOVA provides MS_i, and $\hat{\sigma}^2_{pi,e}$ has been found already in Equation 3.7. So, we need to solve Equation 3.8 for the one unknown in the equation, $\hat{\sigma}^2_i$.

Similarly, we can solve Equation 3.4 for $\hat{\sigma}^2_p$:

$$\hat{\sigma}^2_p = (\text{MS}_p - \hat{\sigma}^2_{pi,e}) / n_i \quad [3.9]$$

Table 3.2 presents the results for the science achievement data from Table 1.1. The computer program BMDP 8V was used to run the ANOVA. (The computer program setup is given in Appendix 3.1.) BMDP 8V gives the sums of squares, degrees of freedom, mean squares, and estimated variance components. So, in most cases, the estimated variance components need not be calculated by hand.[1] Nevertheless, you would get the same values using Equations 3.7 to 3.9.

Interpretation of Variance Components

The estimated variance components from a generalizability study reflect the magnitude of error in generalizing from a person's score on a single item to his or her universe score (the person's average over all items in the universe). These estimated variance components are not the error in generalizing from the test score (average or sum of n_i items on the test).

Consider, for example, the variance component for items. It shows how different the mean (over persons) of any one randomly selected item is expected to be from the mean over all items in the universe. Recall that a variance is defined as the average squared deviation of each point about the mean of the distribution. The variance, then, reflects the squared distance of a single (idealized or average) point from the mean. The estimated variance component for items has an analogous interpretation—the squared distance of a single (idealized) item from the grand mean. The estimated item variance component is based on the variation of each of the eight item means about the mean of these means—the grand mean (i.e., in our example, the mean of all eight items across 20 persons). It indexes the squared distance of one idealized item from the grand mean. Estimated variance components depend on the scale of measurement used. The item variance component, then, reflects the variance of a one-item test. The variance components in Table 3.2 are less than 1 because we are working with variances of 0 or 1 item scores; the maximum variance for 0/1 scores is 0.25. If the items were scored on a different scale, the estimated variance components also would be on a different scale. Consequently, we interpret variance components by their *relative* magnitudes.

As a heuristic in interpreting the relative magnitude of estimated variance components, we can take the sum of the variance components (we call it the *total variance*) and create percentages of this sum that each estimated variance component accounts for. So Table 3.2 gives the percentages, as well as the estimated variance components.

The variance component for persons (the estimated universe-score variance) accounts for only 12% of the total variance. A test that is one item in length, as might be expected, does not do a good job of reliably distinguishing among individuals' science knowledge. The universe-score variance component can be understood better by looking at the mean score, averaging over the eight items, for each person (see Table 1.1, person mean). The scores ranged from 0.125 (only 1 out of 8 items

TABLE 3.3 Frequency Distribution of Subjects' Mean Scores on the Science Achievement Test ($n_i = 8$)

Achievement Test Score	*Frequency*
0.000	0
0.125	3
0.250	2
0.375	6
0.500	6
0.625	1
0.750	0
0.875	0
1.000	2

correct) to 1.000 (all 8 items correct) with a grand mean of 0.4375. The frequency distribution in Table 3.3 succinctly summarizes this information. Most of the scores, however, were tightly clustered around the grand mean, with a few low scores and a few very high scores. This clustering of scores is reflected in the relatively low estimated variance component for universe scores.

The variance component for items (0.0093, about 4% of the total variance; Table 3.2) is small, relative to the other estimated variance components. The small percentage of variation due to items does not mean, however, that differences between items are small in an absolute sense. A way to interpret the variation across items is to use the square root of the variance component for items, that is, the standard deviation. The magnitude of a standard deviation can be used to give a rough approximation of the range of scores in a distribution. In a normal distribution, four standard deviations roughly encompass 95% of the scores; six standard deviations roughly encompass 99% of the scores. The square root of $\hat{\sigma}_i^2 = 0.0093$ is about 0.10. So we would expect the item means to have a range of at least 0.40, if not 0.60, which is a very substantial variation for items scored on a 0–1 scale. In fact, the means of the eight items in the G study range from 0.20 to 0.55, which agrees fairly closely with the predicted range, especially given the small sample of means used here. So items vary considerably in absolute magnitude even though $\hat{\sigma}_i^2$ accounts for only 4% of the total variation.

The largest component in Table 3.2 is the residual—seven times larger than the variance component for universe scores, and much larger than the component for items. The large residual variance suggests (a) a large $p \times i$ interaction, (b) sources of error variability in the science achievement measurement that the one-facet, $p \times i$ measurement has not captured, or (c) both. A large variance component for the $p \times i$ interaction indicates that the relative standing of persons differs from item to item. Variation not captured by the $p \times i$ measurement may be systematic, such as variations in the testing conditions (e.g., time of day), or unsystematic, such as breaking a pencil, or skipping an item partway through the test and marking wrong answers on the rest of the test.

The results of the G study of science achievement make it clear that a one-item test would not provide a good estimate of a person's achievement. In the decision study the decision maker will want to use multiple items. Later chapters show how increasing the number of items will reduce error and increase the generalizability of the scores.

ANOVA Estimation of Variance Components in Multifaceted G Studies

The procedures for estimating and interpreting variance components can be extended easily to multifaceted designs. Consider the two-facet, crossed study of children's help-seeking behavior. On two occasions 13 children were tape-recorded while working on mathematics problems. Two raters coded all children's behavior on both occasions. The design has persons crossed with raters and occasions, denoted as $p \times r \times o$. This design has seven sources of variation and so has seven corresponding variance components that can be estimated and interpreted (see Table 3.4).

Table 3.4 gives the expected mean square equations for the $p \times r \times o$ design. Each mean square from the ANOVA is substituted for its corresponding expected mean square equation to solve for the estimated variance components. The results of the ANOVA appear in Table 3.5. The number of persons (n_p) in this study is 13, the number of raters (n_r) is 2, and the number of occasions (n_o) is 2. These numbers are substituted accordingly into the expected mean square equations. As before, it is easiest to solve the EMS equations from the "bottom up," that is, starting with the equation for the residual first and solving the equation for persons last. The resulting estimated variance components and percentage of total variance that each accounts for also appear in Table 3.5.

TABLE 3.4 Expected Mean Square Equations for the Two-Facet, Crossed $p \times r \times o$ Design

Source of Variation	*Variance Component*	*Expected Mean Square*
Persons (p)	σ^2_p	$\sigma^2_{pro,e} + n_o \sigma^2_{pr} + n_r \sigma^2_{po} + n_r n_o \sigma^2_p$
Raters (r)	σ^2_r	$\sigma^2_{pro,e} + n_p \sigma^2_{ro} + n_o \sigma^2_{pr} + n_p n_o \sigma^2_r$
Occasions (o)	σ^2_o	$\sigma^2_{pro,e} + n_p \sigma^2_{ro} + n_r \sigma^2_{po} + n_p n_r \sigma^2_o$
pr	σ^2_{pr}	$\sigma^2_{pro,e} + n_o \sigma^2_{pr}$
po	σ^2_{po}	$\sigma^2_{pro,e} + n_r \sigma^2_{po}$
ro	σ^2_{ro}	$\sigma^2_{pro,e} + n_p \sigma^2_{ro}$
pro,e (residual)	$\sigma^2_{pro,e}$	$\sigma_{pro,e}$

Several equations are solved here for illustration. The estimated variance component for the residual is simply the mean square for the residual:

$$\hat{\sigma}^2_{pro,e} = \text{MS}_{pro,e} \qquad [3.10]$$

$$= 0.2244$$

TABLE 3.5 Estimated Variance Components for the Behavior Observation Measurement

Source of Variation	*df*	*Mean Square*	*Estimated Variance Component*	*Percentage of Total Variance*[a]
Persons (p)	12	2.5769	0.3974	35
Raters (r)	1	0.6923	0.0096	1
Occasions (o)	1	3.7692	0.1090	10
pr	12	0.3590	0.0673	6
po	12	0.8526	0.3141	28
ro	1	0.3077	0.0064	1
pro,e	12	0.2244	0.2244	20

a. The percentages do not sum exactly to 100 due to rounding.

The estimated residual variance component is substituted into the remaining equations, such as the equation for the $r \times o$ interaction:

$$\hat{\sigma}^2_{ro} = (\text{MS}_{ro} - \hat{\sigma}^2_{pro,e}) / n_p \qquad [3.11]$$

$$= (0.3077 - 0.2244) / 13$$

$$= 0.0064$$

This estimate is substituted into remaining equations that include the $\hat{\sigma}^2_{ro}$ term, such as the equation for occasions:

$$\hat{\sigma}^2_{o} = (\text{MS}_{o} - \hat{\sigma}^2_{pro,e} - n_p\,\hat{\sigma}^2_{ro} - n_r\,\hat{\sigma}^2_{po}) / (n_p\, n_r) \qquad [3.12]$$

$$= [3.7692 - 0.2244 - (13 \times 0.0064) - (2 \times 0.3141)]/(13 \times 2)$$

$$= 0.1090$$

The term $\hat{\sigma}^2_{po}$ would have to be estimated prior to solving this equation.

Interpretation of Estimated Variance Components

The largest variance component, that for persons (0.3974), accounts for about 35% of the total variance in scores. This is the variance component for universe scores and indicates that persons systematically differed in the frequency with which they sought help. The variance component for raters is small, both relative to the other components and in an absolute sense. Using the square root of $\hat{\sigma}^2_r = 0.0096$ (which is about 0.10) as a yardstick, the expected range of rater means is about 0.40 or 0.60. This range is small compared to the range of behavior scores (0 to 4; see Table 1.3). Similarly, the component for the rater-by-occasion interaction is also small. Raters were well calibrated in that they used the same part of the scale and did so consistently at each occasion. The variance component for the interaction between persons and raters (0.0673, 6% of the total variance), in contrast, is somewhat larger. Even though raters used the same standards overall, they disagreed somewhat in the relative standing of persons.

The total variability due to occasions is larger than that due to raters. The variance component for occasions (0.1090, 10% of the total variance) indicates that the overall frequency of children's help-seeking differed somewhat from one occasion to another. The large variance component for the interaction between persons and occasions (0.3141, 28% of the total variance) indicates that the relative standing of persons differed from one occasion to another. The child who sought the most help on one occasion did not necessarily seek the most help on another occasion.

Finally, the variance component for the residual $\hat{\sigma}^2_{pro,e} = 0.2244$ (20% of the total variance) shows that a substantial proportion of the variance was due to the three-way interaction between persons, raters, and occasions and/or other unsystematic or systematic sources of variation that were not measured in this study.

On Deriving Expected Mean Squares Equations

Variance components are estimated from sample data by setting expected mean square equations equal to the sample mean squares from the ANOVA and then solving them for the variance component of interest. Each alternative G-study design has a different set of expected mean square equations, as exemplified by the equations for the random, one- and two-facet, completely crossed measurements presented above. A number of different tools are available for finding the set of equations for a particular design.

First, statistical computer programs such as BMDP's 8V and SAS's VARCOMP provide the appropriate expected mean squares equations for G-study designs. Their printouts contain both the equations and the numerical values of the estimated variance components. Table 3.6 contains the variance component information provided on a BMDP 8V printout for the science achievement data with 20 persons and 8 items. For the expected mean squares equations, the numbers in parentheses refer to the "sources" or "components" of variability in observed scores, and the numbers not in parentheses refer to the number of times each component is sampled.[2] Note that the expected mean squares equations in Table 3.6 correspond to those in Table 3.1, and the estimated variance components in Table 3.6 are the same as those in Table 3.2. A computer program developed specifically for generalizability analyses, GENOVA (Crick & Brennan, 1982), also gives expected mean square equations, as well as estimated variance components and generalizability coefficients, for a wide variety of G-study and D-study designs.

TABLE 3.6 BMDP 8V Printout for Estimated Variance Components Using the Science Achievement Test Data

Source	*Expected Mean Square* [a]	*Estimates of Variance Components*
1 Mean	160(1) + 8(2) + 20(3) + (4)	0.1874
2 Persons	8(2) + (4)	0.0305
3 Items	20(3) + (4)	0.0093
4 *pi*	(4)	0.2103

a. The numbers in parentheses represent the (1) deviation of the grand mean from zero, (2) person variance component, (3) item variance component, and (4) residual variance component. The other numbers represent the appropriate sample size multipliers n_p, n_i, and $n_p\, n_i$.

Second, algorithms, such as the Cornfield-Tukey (1956) or Millman-Glass (1967) algorithm (e.g., Kirk, 1982), provide a method for deriving expected mean square equations. These algorithms are tedious, however, so we do not present them here.

Third, a statistics book may be used as a source of expected mean square equations. Looking them up is easier than using a computer program before you have collected data, and it is always easier than working through an algorithm. In Appendices 3.2 and 4.3 (for chapter 4) we provide the expected mean square equations for a variety of two-facet G-study designs (see also Brennan, 1983; Cronbach et al., 1972; Kirk,1982).

Negative Estimates of Variance Components

On occasion, a sample estimate of a variance component may be negative. To see how this can happen, suppose that a persons-by-items ANOVA ($n_p = 40$, $n_i = 20$) returned the following mean squares:

$$MS_p = 22.50$$

$$MS_i = 5.91$$

$$MS_{pi,e} = 6.25$$

If the estimated variance component for items is calculated using Equation 3.8, the estimate is very close to zero, but negative:

$$\hat{\sigma}_i^2 = (\text{MS}_i - \hat{\sigma}_{pi,e}^2) / n_p$$

$$= (5.91 - 6.25) / 40$$

$$= -0.0085$$

Negative estimates can arise because of a misspecification of the measurement model or because of sampling error. When negative estimates are large in relative magnitude, chances are that they reflect a misspecification of the measurement model. For example, in a study of military job performance measurements, two facets were the job-task (e.g., radio operator, map reading) and "stations" (job-tasks were grouped into "stations" for administrative convenience). So tasks were nested within stations. At some stations the tasks addressed a similar job area, such as land navigation; at other stations the tasks were unrelated to one another. Marines performed the set of job-tasks at one station and then moved to the next station. Shavelson, Mayberry, Li, and Webb (1990) found large negative variance components associated with the stations facet, and its interaction with other facets. Variation among tasks within stations was often greater than variation between stations, resulting in large negative estimates of variance components for effects concerning stations. By respecifying the measurement model, in this case dropping the stations facet from the model, the large negative variance components were eliminated. When negative estimates of variance components arise out of model misspecification, thought should be given to a better specification of the measurement. No doctoring of the negative estimates will make sense.

When negative estimates are small in relative magnitude (near zero), they probably arise because a small sample was drawn from an indefinitely large universe (sampling error). Something must be done because negative variances are conceptually impossible. One approach, due to Cronbach et al. (1972), is to set the negative estimate to zero. If that variance component appears in expected mean square equations for other variance components, a value of zero, rather than the negative estimate, is used to calculate the new variance component estimates. For example, from Table 3.4 we see that the variance component for the person-by-occasion interaction is involved in two other expected mean squares equations—the equations for the occasion component and the person component. Following Cronbach et al. (1972), if the numerical estimate of σ^2_{po} were negative, it would be set to 0, and that value would be used in solving the equations in Table 3.4 for $\hat{\sigma}^2_o$ and $\hat{\sigma}^2_p$.

Another approach (see Brennan, 1983) also sets the negative estimate to zero, but it uses the original negative estimate in the calculation of all other variance components. This procedure amounts to using the estimated variance components, regardless of whether one (or more) is negative, in calculations of other variance components. Once all variance components have been estimated, the negative component(s) is (are) reported as zero.

The strength of one approach is the weakness of the other. The first immediately gets rid of a theoretically impossible nuisance and does not deign to use the negative estimate in further calculations. But this approach produces biased estimates of variance components because they were calculated with the negative estimate set to zero. The second approach, although uncomfortably using the negative estimate throughout the variance component calculations, returns unbiased estimates of the variance components. The bottom line is that either method is used commonly; neither is satisfying. Some relatively new estimation techniques have been developed that are considerably more elegant. Presentation of them, however, is beyond the purview of this *Primer* (see Shavelson, Webb, & Rowley, 1989).

Summary

The analysis of variance (ANOVA) partitions the total variability in a data set into its separate sources of variation. Applied to G theory, the ANOVA partitions the total variability in a behavioral measurement into separate sources for the object of measurement (usually persons), each other source (facets), and all the interactions that can be estimated. For the science achievement test example, the ANOVA partitions the total variability among item scores into effects for persons, items, and the residual (persons-by-items interaction, and unmeasured effects). Once the variability among scores has been partitioned, the observed mean squares from the ANOVA can be set equal to the expected mean squares from G theory. Then, the variance components can be estimated by solving for the variance component of interest in each equation. In the science achievement test example, by setting the ANOVA mean square for persons equal to the expected mean square for persons and solving for the person variance component, the universe-score variance, σ^2_p, can be estimated by $\hat{\sigma}_p^2$. Similarly, by setting the mean square for items equal to its expectation, the variance component for items, σ_i^2, can be estimated by $\hat{\sigma}_i^2$. And by setting the residual mean square equal to its expectation, the variance component for the residual, $\sigma_{pi,e}^2$, can be estimated by $\hat{\sigma}_{pi,e}^2$. This procedure also applies to multifaceted G studies. By setting the ANOVA mean squares equal to their corresponding expected mean squares and solving the set of equations, variance components can be estimated from G-study data.

In the science achievement test example, the estimated variance component for persons reflects systematic variability among students' science knowledge, the universe-score variance (cf. true-score variance in classical theory). The variance component for items reflects differences in item difficulties. The residual variance component reflects (a) the fact that some items are easier for some people while other items are easier for some other people, and/or (b) other unidentified sources of systematic and random variation.

Unlike correlations, variance components are not measured on a universal metric. Hence, their interpretation depends on the relative magnitude of the variance components. One heuristic way to index relative magnitude is to calculate the percentage of total variance that each component accounts for.

For example, in the science achievement G study, the total variance was $\hat{\sigma}_p^2 + \hat{\sigma}_i^2 + \hat{\sigma}_{pi,e}^2 = 0.2501$. The estimated person variance accounted for about 12% of the total; item variance accounted for about 4%, and the residual accounted for the remainder, 84%. Either there was a tremendous persons-by-items interaction, or the G-study design ignored some other, important facet of the science achievement measurement, or there was substantial random variation in the measurement (or some combination of these factors).

Once in a while an estimate of a variance component is negative—a theoretical impossibility (a variance cannot be negative). This nevertheless may arise from model misspecification or sampling error. If the former, the G-study model should be respecified and variance components reestimated. If the latter, two methods are presented that converge in reporting negative estimates as zero.

Appendix 3.1

BMDP 8V Setup for G Study of Data in Table 1.1

```
/PROBLEM     TITLE='SAGE MONOGRAPH: PERSONS × ITEMS G
                  STUDY DESIGN'.
/INPUT       VARIABLES ARE 8.
             FORMAT IS FREE.
/VARIABLE    NAMES ARE ITEM1, ITEM2, ITEM3, ITEM4, ITEM5,
             ITEM6, ITEM7, ITEM8.
/DESIGN      LEVELS ARE 20,8.
             NAMES ARE PERSONS,ITEMS.
             RANDOM ARE PERSONS,ITEMS.
             MODEL IS 'P,I'.
             PRINT IS PI.
/PRINT       LINE=60.
/END
0 1 0 0 0 1 0 1
1 0 1 0 0 0 0 1
1 1 1 0 0 0 0 0
1 1 0 0 1 0 0 1
1 1 1 1 0 0 0 1
1 1 1 1 1 1 1 1
1 0 1 0 0 1 1 0
1 0 1 0 0 0 1 1
1 0 0 0 0 1 1 1
0 1 0 0 1 1 0 0
0 0 0 1 1 1 0 0
0 0 1 0 0 0 0 0
1 1 1 1 1 1 1 1
0 0 0 0 0 1 1 1
0 0 1 0 0 0 0 1
1 1 1 0 0 1 0 0
0 1 0 0 0 0 0 0
1 0 0 0 0 1 1 1
0 0 0 0 0 1 1 0
0 1 0 0 0 0 0 0
```

Appendix 3.2

Expected Mean Squares Equations for a Crossed, Two-Facet Random-Effects Design

$p \times i \times j$ Design

$$\text{EMS}_p = \sigma^2_{pij,e} + n_i\sigma^2_{pj} + n_j\sigma^2_{pi} + n_i n_j \sigma^2_p$$

$$\text{EMS}_i = \sigma^2_{pij,e} + n_j\sigma^2_{pi} + n_p\sigma^2_{ij} + n_p n_j \sigma^2_i$$

$$\text{EMS}_j = \sigma^2_{pij,e} + n_i\sigma^2_{pj} + n_p\sigma^2_{ij} + n_p n_i \sigma^2_j$$

$$\text{EMS}_{pi} = \sigma^2_{pij,e} + n_j\sigma^2_{pi}$$

$$\text{EMS}_{pj} = \sigma^2_{pij,e} + n_i\sigma^2_{pj}$$

$$\text{EMS}_{ij} = \sigma^2_{pij,e} + n_p\sigma^2_{ij}$$

$$\text{EMS}_{pij,e} = \sigma^2_{pij,e}$$

Exercises

1. As part of an experiment on person perception and attribution, 89 people were given information about a fictitious person and then were administered a 15-item test to measure their recall about the fictitious person. In Table 3.7 are the results of a two-way ANOVA.
   a. Calculate the relevant variance components and the percentage of variance accounted for by each.
   b. Interpret the estimated variance components.
   c. The F values for the main effects of persons and items are both statistically significant. Of what use is this information in G theory?
   d. The estimated variance components in (b) concern the generalizability of what length test (i.e., how many items)?

TABLE 3.7

Source of Variation	*Sum of Squares*	*df*	*Mean Square*	*F*
Persons (p)	47.491	88	0.540	3.15*
Items (i)	65.263	14	4.662	27.22*
pi,e	211.004	1232	0.171	—

*$p < .01$

2. A persons-by-raters design has two variance components that concern raters: σ_r^2 and $\sigma_{pr,e}^2$.
   a. Describing both as "differences between raters" is ambiguous. Give more precise descriptions.
   b. If σ_r^2 is large, must $\sigma_{pr,e}^2$ also be large?
3. In a hypothetical G study of vocational interest, six persons completed five items representing clerical interest on two occasions. The data are presented in Table 3.8.
   a. Run a three-way ANOVA to get the mean squares.
   b. List the expected mean square equations.
   c. Calculate the estimated variance components.
   d. Explain Cronbach's and Brennan's methods for handling negative variance components.
   e. Apply both methods to the data in this exercise.
   f. Interpret the estimated variance components.

TABLE 3.8

		Item				
Person	*Occasion*	*1*	*2*	*3*	*4*	*5*
1	1	5	5	4	4	3
	2	5	5	4	4	4
2	1	3	4	2	3	2
	2	5	5	2	2	4
3	1	1	4	5	2	4
	2	2	3	5	2	3
4	1	3	3	2	2	4
	2	4	4	2	2	2
5	1	1	4	4	2	2
	2	2	4	4	3	5
6	1	2	2	2	1	2
	2	1	3	1	1	2

Source: The data are from "Using Generalizability Theory in Counseling and Development" by N. M. Webb, G. L. Rowley, and R. J. Shavelson, 1988, *Measurement and Evaluation in Counseling and Development, 18.*

Answers to Exercises

1. a. Persons: 0.025 (10%); Items: 0.050 (20%); *pi,e:* 0.171 (70%).
   b. The variance component for persons (universe-score variance) is relatively small, suggesting that persons did not differ much in their level of recall. The variance component for items is relatively large, suggesting that some items were more difficult than others, averaging over all persons. The residual term (*pi,e*) is very large, due to (a) persons differing on which items they found difficult, (b) unmeasured sources of variation (systematic or unsystematic), or (c) both.
   c. None. G theory interprets the magnitudes of estimated variance components and does not use the information from inferential tests.
   d. The estimated variance components in (b) concern the generalizability of a *single item, not* a 15-item test such as that used to collect data in the G study. In this case the large estimated variance components for items and the residual show that generalizing from a single-item test to the universe of items would be hazardous.

TABLE 3.9

Source of Variation	*Sum of Squares*	*df*	*Mean Square*
Persons (*p*)	34.7333	5	6.9467
Items (*i*)	14.1000	4	3.5250
Occasions (*o*)	1.0667	1	1.0667
pi	30.1000	20	1.5050
po	3.3333	5	0.6667
io	1.4333	4	0.3583
pio,e	11.1667	20	0.5583

2. a. σ_r^2 concerns main-effect differences among raters, whether raters use different scales averaging over persons, that is, whether some raters "see" more behavior than other raters do. $\sigma_{pr,e}^2$ concerns the interaction between persons and raters (confounded with unmeasured error), whether the relative standing of persons differs across raters.

   b. No. σ_r^2 and $\sigma_{pr,e}^2$ are independent. Both can be large, both can be small, and one can be large while the other is small.

3. a. See Table 3.9.

   b. EMS $(p) = \sigma_{pio,e}^2 + 2\sigma_{pi}^2 + 5\sigma_{po}^2 + 10\sigma_p^2$

   EMS $(i) = \sigma_{pio,e}^2 + 2\sigma_{pi}^2 + 6\sigma_{io}^2 + 12\sigma_i^2$

   EMS $(o) = \sigma_{pio,e}^2 + 5\sigma_{po}^2 + 6\sigma_{io}^2 + 30\sigma_o^2$

   EMS $(pi) = \sigma_{pio,e}^2 + 2\sigma_{pi}^2$

   EMS $(po) = \sigma_{pio,e}^2 + 5\sigma_{po}^2$

   EMS $(io) = \sigma_{pio,e}^2 + 6\sigma_{io}^2$

   EMS $(pio,e) = \sigma_{pio,e}^2$

   c. See Table 3.10.

   d. Cronbach's method sets negative values equal to zero and uses zero in all calculations involving those negative values. Brennan's method sets negative values equal to zero but uses the negative values in all calculations.

   e. The relatively large variance component for persons shows that, averaging over occasions and items, persons in the sample differ in their clerical interest. The moderately large variance component for the main effect of items shows that some items reflected more clerical interest in the whole group than did other items. The large persons-by-items interaction shows that the relative standing of persons on clerical

TABLE 3.10

Source of Variation	*Variance Components*	*Cronbach's Method*	*Brennan's Method*
Persons (p)	0.5333	0.5333	0.5333
Items (i)	0.1850	0.1683	0.1850
Occasions (o)	0.0200	0.0133	0.0200
pi	0.4733	0.4733	0.4733
po	0.0217	0.0217	0.0217
io	–0.0333	0	0
pio,e	0.5583	0.5583	0.5583

interest differed across items. The small variance component for the persons-by-occasions interaction shows that the relative standing of persons on clerical interest remained stable from one occasion to the next. The negative effect, set to zero, for the items-by-occasions shows that mean item scores had the same relative standing across occasions. The large residual effect suggests a large persons-by-items-by-occasions interaction, unmeasured sources of variation, or both. Overall, more of the variability comes from items than from occasions.

Notes

1. The exception occurs when estimated variance components are negative. How to deal with negative estimated variance components is explored later in this chapter.

2. Although BMDP 8V provides a variance component for the grand mean, in G theory it is ignored. This variance component indexes the difference between the grand mean and zero, a characteristic of the metric chosen. Consequently, the information it provides is not of particular interest in G theory.

4

Generalizability Studies with Nested Facets

Chapter 3 showed how to use the analysis of variance to estimate variance components from random-effects, one- and two-facet crossed designs. This chapter develops the measurement model for generalizability studies with nested facets.

The measurements considered so far have involved completely crossed designs. In the science achievement test study with a $p \times i$ design, all persons were administered all items. In the behavior observation study with a $p \times r \times o$ design, all persons were rated by both raters on both occasions. This need not be the case. The investigator could have used different raters on different occasions. In such a design, we speak of raters nested within occasions; they are not crossed with occasions.

Definition

In the analysis of variance, one factor (call it A) is *nested* within another (call it B) if (a) multiple levels of A are associated with each level of B, and (b) different levels of A are associated with each level of B. In G theory, nested facets are defined in the same way as in the analysis of variance. An achievement test with several subtests, for example, has different sets of items (i) associated with each subtest (s). So items are *nested* within subtests. We use the notation $i{:}s$ or $i(s)$ to denote facet i nested within facet s. Both conditions a and b must be satisfied for one facet to be nested within another. If only the first condition were satisfied—multiple levels of A associated with each level of B but the same levels of A are associated with each level of B—the facets would be crossed. If only the second condition were satisfied—a single, different level of A for each level of B—the design would have A *confounded* with B, not A nested within B.

Schematic (a) in Figure 4.1 shows a completely crossed, persons-by-raters-by-occasions measurement. This is the design of the behavior observation measurement used throughout much of the *Primer* to exemplify a two-facet, crossed design. Schematic (b) shows a nested measurement with raters 1 and 2 coding behavior on occasion 1, and two different raters, 3 and 4, coding behavior on occasion 2. Multiple conditions of the rater facet and different conditions of the rater facet are associated with each condition of the occasion facet. This measurement has raters nested within occasions, but raters and occasions are both crossed with persons. Schematic (c) shows the rater facet confounded with the occasion facet. A different condition of the rater facet is associated with each condition of the occasion facet. We do not consider confounded measurements in this *Primer*.

G studies usually have naturally nested facets, but they also may have nested facets by choice. In a "naturally nested" setting, the choice about whether to nest facets is not available. They occur that way by definition. The universe of admissible observations is nested. The achievement measure with subtests, mentioned above, is one example. The items associated with one subtest (say, reading comprehension) are part of that subtest and no other (say, arithmetic computation). By definition, items are nested within subtests.

Sometimes, facets are nested in G studies because of cost or logistic considerations; that is, the universe of admissible observations is crossed, but the investigator may choose to have a nested rather than a crossed measurement. Consider a measurement in which employees at a company are to be evaluated on three occasions by two observers. In the universe of admissible observations, each employee could, theoretically, be observed on all occasions by all observers (a crossed universe). The design of the decision study could have employees crossed with occasions and observers: Employees may be evaluated on the same three occasions by the same two observers. Because of logistics, however, a different pair of observers may be present on each of the three occasions. In this design, observers are nested within occasions.

In the examples just described, one facet was nested within another facet (e.g., items nested within subtests). A facet may be nested also within the object of measurement. For example, raters could be nested within employees; different raters evaluate each employee. It is also possible for the object of measurement to be nested within a facet. For

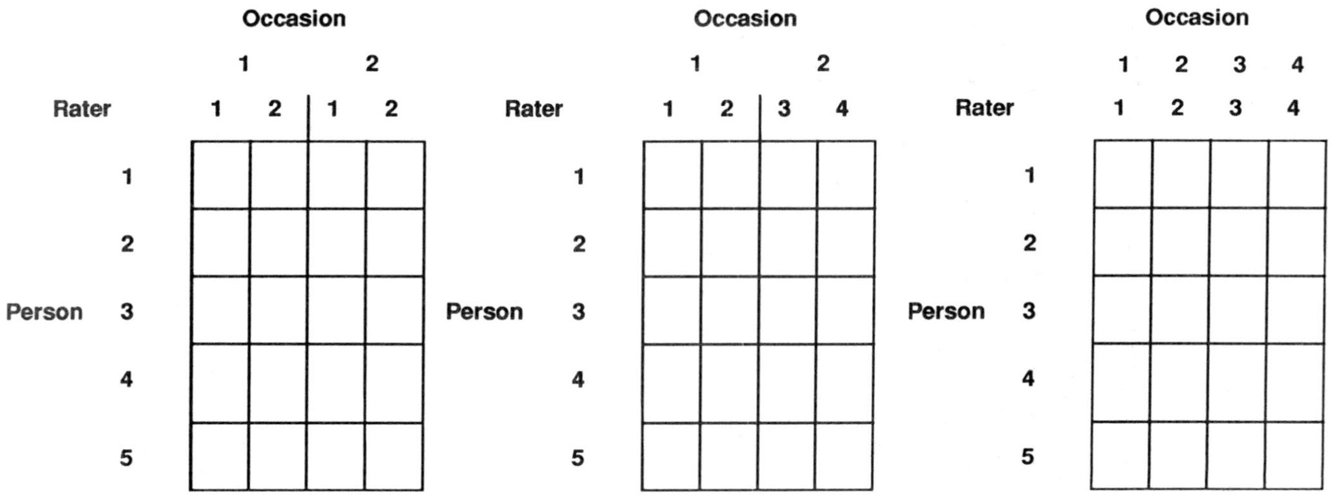

Occasion
1 2
Rater 1 2 1 2
Person 1 2 3 4 5

Occasion
1 2
Rater 1 2 3 4
Person 1 2 3 4 5

Occasion
1 2 3 4
Rater 1 2 3 4
Person 1 2 3 4 5

Figure 4.1. Schematics of (a) Crossed, (b) Nested, and (c) Confounded Designs

example, employees could be nested within raters. Procedures are available for estimating variance components in this case (see Shavelson et al., 1990), but because they are complicated, we do not consider such designs in this *Primer.*

Formal Model: One-Facet Nested Design

To develop the formal model for the one-facet, nested measurement, consider a study of mathematics achievement. The intent is to administer a large sample of items from the universe, a set too large for any one student to tolerate. So, each student receives a random sample of 20 items; no two students receive the same items. Items (i), then, are nested within persons (p). In notational form, the design is $i{:}p$ or $i(p)$.

For an $i{:}p$ design, an observed score for one person on one item (X_{pi}) can be decomposed into the following effects:

$$\begin{aligned} X_{pi} = \; & \mu && \text{[grand mean]} \\ & + (\mu_p - \mu) && \text{[person effect]} \\ & + (X_{pi} - \mu_p) && \text{[residual effect]} \end{aligned} \qquad [4.1]$$

Unlike the one-facet crossed design ($p \times i$), the nested $i{:}p$ design has no separate term for the item effect. The item effect is part of the residual term. This occurs because μ_i and μ_{pi} are confounded. Because different persons are administered different items, the item effect cannot be estimated independently of the person-by-item interaction. The full form of the residual effect shows the item effect as part of the residual (based on Brennan, 1983, p. 22):

$$\begin{aligned} (X_{pi} - \mu_p) &= (X_{pi} - \mu_p) + (\mu_i - \mu_i) + (\mu - \mu) \\ &= (\mu_i - \mu) + (X_{pi} - \mu_i - \mu_p + \mu) \end{aligned} \qquad [4.2]$$

Each effect in Equation 4.1, other than the grand mean, has a distribution with mean zero and an associated variance. The variance component for persons is the universe-score variance and is defined in the same way as for crossed designs:

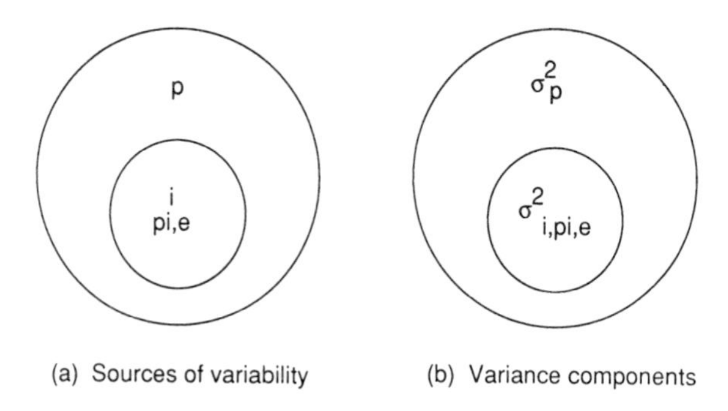

Figure 4.2. Venn Diagrams for a One-Facet, Nested *i:p* Design

$$\sigma_p^2 = \mathrm{E}_p\,(\mu_p - \mu)^2 \qquad [4.3]$$

The variance component for the residual is

$$\sigma_{i,pi,e}^2 = \mathrm{E}_p \mathrm{E}_i\,(X_{pi} - \mu_p)^2 \qquad [4.4]$$

The notation for the residual variance component ($\sigma_{i,pi,e}^2$) reflects the fact that the item effect is confounded with the effect for the interaction between persons and items (which is confounded with unsystematic or unmeasured sources of variation).

The variance of the collection of observed scores X_{pi} for all persons and items is the sum of the two variance components:

$$\sigma^2\,(X_{pi}) = \sigma_p^2 + \sigma_{i,pi,e}^2 \qquad [4.5]$$

The Venn diagrams showing the sources of variance in the one-facet nested design appear in Figure 4.2. Numerical estimates of the variance components come from the analysis of variance for a nested design (developed in the next section).

The one-facet, nested design illustrates the disadvantage of using a nested design, as opposed to a crossed design, in a G study. The drawback stems from the fact that it is not possible to obtain a separate estimate of the item main effect. For a G study, it is desirable to use as

TABLE 4.1 Variance Components for One-Facet, Nested, Random G Study of Mathematics Achievement

Source of Variation	*Variance Component*	*Estimated Variance Component*	*Percentage of Total Variance*
Persons (p)	σ_p^2	0.0750	27
Items:Persons (i:p)	$\sigma_{i,pi,e}^2$	0.2000	73

fully crossed a design as possible to estimate as many sources of variability in the measurement as possible (Cronbach et al., 1972). This statement applies, of course, to situations where the investigator has a choice between crossing and nesting in the design. Where there is nesting in the universe of admissible observations (such as the example of items nested within subtests in the previous section), the investigator has no choice but to incorporate that nesting into the generalizability study.

Numerical Example of One-Facet, Nested Design

Consider a hypothetical study of mathematics achievement in which five persons (p) each take a different pair of items (i). This design has items nested within persons, denoted as i:p or $i(p)$. The data for this hypothetical study and the computer analysis setup using BMDP 8V appear in Appendix 4.1. The resulting variance components appear in Table 4.1.

As can be seen in Table 4.1, most of the variance in mathematics scores is attributable to the residual variance component ($\sigma_{i,pi,e}^2$ = 0.2000; 73% of the total variance). In this residual term the item effect, the person-by-item interaction, and remaining sources of systematic and unsystematic variation not measured in the study are all confounded. Because different persons received different items, it is impossible to determine whether items differed in difficulty level (item effect), or whether some items were particularly difficult for some students (person-by-item interaction effect). Particularly low performance on an item that a person received could have arisen because the item was especially difficult or because it was especially difficult for that particular person. The low performance could have also resulted from remaining unmeasured sources of variation, such as the person accidentally marking the wrong answer on the answer sheet.

These results show that a nested design leaves important questions unanswered; in this study no unambiguous conclusion can be drawn about item difficulty. A crossed $p \times i$ generalizability study, in which all persons take the same items, would yield interpretable information about item difficulty.

Two-Facet Nested Designs

To develop the notion of a two-facet, nested measurement, consider a study of teacher behavior (Erlich & Shavelson, 1978) in which five teachers were videotaped on three occasions and three raters coded each of the 15 videotapes. In this study, the occasions were different for each teacher. The object of measurement is persons (p), the teachers. The facets are occasions (o) and raters (r). In this design the occasion facet is nested within teachers because there are multiple occasions per teacher and the occasions differ from teacher to teacher. The rater facet is crossed with both occasions and teachers because each rater coded all teachers on all occasions. We denote this design as $(o{:}p) \times r$. This is a "partially" nested design because it has both crossed and nested effects. "Fully" nested designs have no crossed effects in the design; all effects are nested. Appendix 4.2 summarizes all partially and fully nested designs with two facets.

Equations 4.6 and 4.7 and the Venn diagrams in Figure 4.3 show the decomposition of score effects and the variance components for the $(o{:}p) \times r$ design. Compared with a two-facet crossed design, some sources of variability cannot be estimated separately in this nested design because of confounding. The occasion effect (o) is confounded with the person-by-occasion interaction (po). So the variance component for occasions (σ^2_o) is confounded with the variance component for the person-by-occasion interaction (σ^2_{po}). Hence, the variance component for these confounded effects is denoted as $\sigma^2_{o,po}$. The rater-by-occasion interaction (ro) is confounded with the three-way interaction and remaining unmeasured sources of error (pro,e). So the combined variance component is $\sigma^2_{ro,pro,e}$. The $(o{:}p) \times r$ design, then, has five

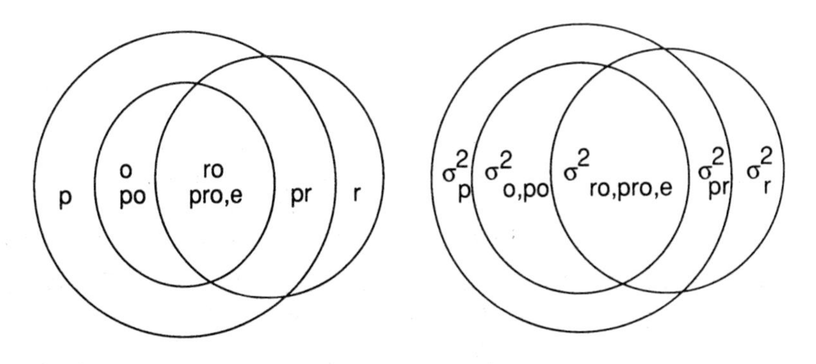

Figure 4.3. Venn Diagrams for the $(o{:}p) \times r$ Design

variance components that can be estimated separately, compared with seven in a two-facet crossed design.

$$\begin{aligned} X_{pro} = {} & \mu \\ & + \mu_p - \mu \\ & + \mu_r - \mu \\ & + \mu_{po} - \mu_p \\ & + \mu_{pr} - \mu_p - \mu_r + \mu \\ & + X_{pro} - \mu_{pr} - \mu_{po} + \mu_p \end{aligned} \quad [4.6]$$

$$\sigma^2 (X_{pro}) = \sigma^2_p + \sigma^2_r + \sigma^2_{o,po} + \sigma^2_{pr} + \sigma^2_{ro,pro,e} \quad [4.7]$$

For purposes of this illustration, we consider persons, raters, and occasions to be random. That is, we assume that the raters and occasions used in this study were selected at random from an indefinitely large universe of raters and occasions, or at least can be considered exchangeable with any of the other raters and occasions in the universe (see chapter 1 for a discussion of exchangeability).

Table 4.2 presents the estimated variance components from the analysis of the five teachers using as the dependent variable, "direction to student to try again." (Expected mean squares equations for two-facet, nested designs are given in Appendix 4.3.) The variance component for

TABLE 4.2 Two-Facet, Partially Nested Random G Study of Teacher Behavior with $(o{:}p) \times r$ Design

Source of Variation	*df*	*Mean Squares*	*Variance Component*	*Estimated Variance Component*	*Percentage of Total Variance*
Teachers (*p*)	4	97.530	σ_p^2	8.288	32
Rater (*r*)	2	52.878	σ_r^2	2.613	10
Occasions:Teachers (*o:p*)	10	13.683	$\sigma_{o,po}^2$	4.622	18
pr	8	22.938	σ_{pr}^2	1.537	6
o:pr,e	20	9.072	$\sigma_{ro,pro,e}^2$	9.072	35

persons (σ_p^2) is the universe-score variance; it shows the amount of systematic variability between teachers in their behavior. The estimated variance component is substantial ($\sigma_p^2 = 8.288$; 32% of the total variance); it shows that teachers differed in their behavior.

The substantial variance component for raters ($\sigma_r^2 = 2.613$; 10% of the total variance) indicates that raters differed in how much behavior they "saw," averaging over teachers and occasions.

Since occasions are nested within teachers, it is impossible to separate the occasion main effect from the interaction between teachers and occasions. We interpret the substantial variance component for those combined effects ($\sigma_{o,po}^2 = 4.622$; 18% of the total variance) as indicating that teacher behavior differed from one occasion to another. We do not know whether one occasion produced more behavior than another (occasion main effect), whether the relative standing of teachers differed from one occasion to another (teacher-by-occasion interaction), or both.

The nonnegligible variance component for the interaction between persons and raters ($\sigma_{pr}^2 = 1.537$; 6% of the total variance) suggests that the relative standing of teachers differed somewhat from one rater to another.

The interaction between raters and occasions, the three-way interaction between persons, raters, and occasions and unmeasured variation are confounded in this design. The large residual component ($\sigma_{ro,pro,e}^2 = 9.072$; 35% of the total variance) indicates that a substantial amount of variation is due to these confounded sources of variation.

Summary

Not all measurements involve crossed designs in which, for example, all students take the same items on an achievement test or all children are observed by the same raters on the same occasions. Rather, some achievement measurements use different items for different students, thereby increasing the sample of items from the universe. And some observation measurements use different raters at each occasion, thereby easing logistical problems. In these latter cases, we speak of nested measurements. Facet A is said to be nested within facet B if (a) two or more conditions of A are observed with each condition of B, and (b) different levels of A are associated with each level of B. In the achievement example, different sets of items would be associated with different students.

With nested facets, it is not possible to estimate all variance components separately. For example, the model for a measurement with each student receiving a different set of items (i.e., items nested within student) would contain three terms: the constant (μ), the person effect ($\mu_p - \mu$), and a residual effect ($X_{pi} - \mu_p$). Unlike the crossed $p \times i$ design in which all students take the same items, the item effect in this nested design cannot be estimated separately from the person-by-item interaction because of missing cells in the design (e.g., person 1 is missing data on items person 2 received). Correspondingly, only two variance components can be defined for the $i{:}p$ design: σ_p^2 and $\sigma_{i,pi,e}^2$. As with crossed measurements, the ANOVA is used to estimate the variance components in a nested design.

Multifaceted measurements also may have one or more nested facets. For example, a whole family of two-facet, nested designs exists. Consider, for instance, ratings of teacher behavior on multiple occasions by different raters. The two facets are raters (r) and occasions (o). Teachers were videotaped on different occasions. We write this design as $(o{:}p) \times r$—occasions nested within persons, and raters crossed with occasions and persons. With this measurement model, variance components can be estimated for persons, raters, occasions nested within persons, persons-by-raters, and a residual: σ_p^2, σ_r^2, $\sigma_{o,po}^2$, σ_{pr}^2, and $\sigma_{ro,pro,e}^2$. Compared with a completely crossed design, fewer variance components can be estimated with this nested design.

The goal of G studies is to estimate as many variance components as possible. Consequently, crossed facets should be preferred over nested facets wherever possible. For example, given the trade-off of a persons-by-items design ($p \times i$) and an items-nested-within-person design ($i{:}p$), the former is preferred. When one facet is naturally nested in another, however, such as items nested within subtests, the only option is to have nesting in a G study.

Appendix 4.1

BMDP 8V Setup: One-Facet, Nested *i:p* Design

```
/PROBLEM       Title= 'I:P DESIGN'.
/INPUT         VARIABLES ARE 2.
               FORMAT IS FREE.
/DESIGN        NAMES ARE PERSONS, ITEMS.
               LEVELS ARE 5,2.
               RANDOM ARE PERSONS, ITEMS.
               MODEL IS 'P, I(P)'.
/PRINT         LINE=60.
/END
0 0
1 0
0 1
1 1
1 1
```

Appendix 4.2

Alternative Nested, Two-Facet, Random Designs

The five major types of nested designs with two facets are

1. $p \times (i{:}j)$	or	$p \times (j{:}i)$
2. $(i{:}p) \times j$	or	$(j{:}p) \times i$
3. $i{:}(p \times j)$	or	$j{:}(p \times i)$
4. $(i \times j){:}p$		same as $(j \times i){:}p$
5. $i{:}j{:}p$	or	$j{:}i{:}p$

Additional designs are possible with the object of measurement (p) nested within one or more facets, but they are not treated in this book because they complicate the estimation of universe-score variance (see Shavelson et al., 1990, for estimation procedures for such designs).

Parts 1 through 5 of this appendix summarize the decomposition of effects in each design. For the description of each design, we use the hypothetical example of a mathematical reasoning test in which all items (i) have an open-ended format (as opposed to multiple choice). All item responses are graded by multiple judges (j). The object of measurement is persons (p) who are administered the test.

Part 1: Decomposition of Score Effects and Variance for the $p \times (i{:}j)$ Design

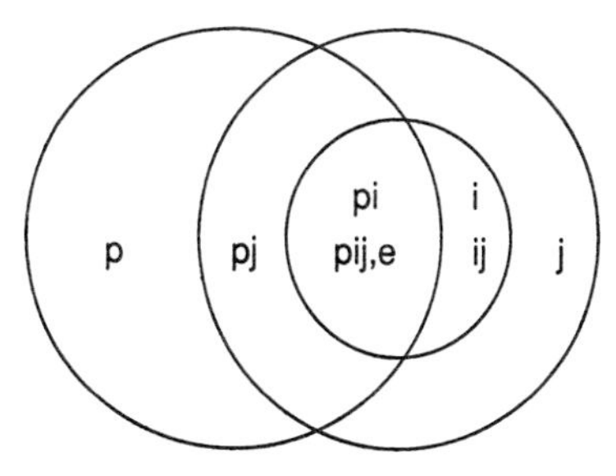

Figure 4.4. Sources of Variability

$$\begin{aligned} X_{pij} = \; & \mu \\ & + (\mu_p - \mu) \\ & + (\mu_j - \mu) \\ & + (\mu_{ij} - \mu_j) \\ & + (\mu_{pj} - \mu_p - \mu_j + \mu) \\ & + (X_{pij} - \mu_{pj} - \mu_{ij} + \mu_j) \end{aligned}$$

$$\sigma^2 (X_{pij}) = \sigma_p^2 + \sigma_j^2 + \sigma_{i,ij}^2 + \sigma_{pj}^2 + \sigma_{pi,pij,e}^2$$

Description: If the test consists of 20 items, four judges may each evaluate five different items (judge 1 evaluates items 1 through 5; judge 2 evaluates items 6 through 10, and so on). Items are nested within judges because there are multiple items per judge and the items for each judge are different. The item and judge facets are both crossed with persons because each person takes all items and each person is evaluated by each judge.

Part 2: Decomposition of Score Effects and Variance for the $(i{:}p) \times j$ Design

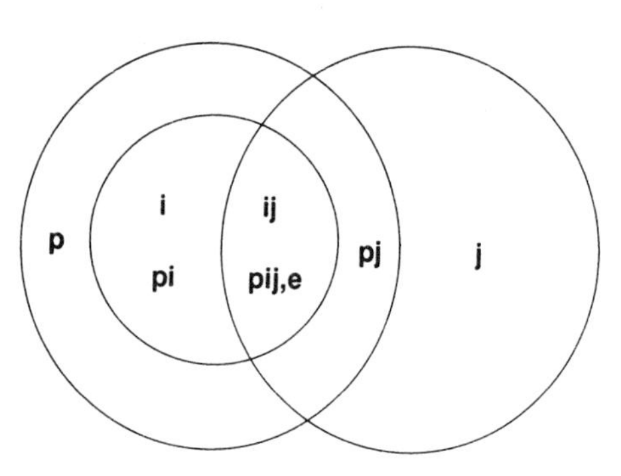

Figure 4.5. Sources of Variability

$$\begin{aligned} X_{pij} = \; & \mu \\ & + (\mu_p - \mu) \\ & + (\mu_j - \mu) \\ & + (\mu_{pi} - \mu_p) \\ & + (\mu_{pj} - \mu_p - \mu_j + \mu) \\ & + (X_{pij} - \mu_{pj} - \mu_{pi} + \mu_p) \end{aligned}$$

$$\sigma^2 (X_{pij}) = \sigma_p^2 + \sigma_j^2 + \sigma_{i,pi}^2 + \sigma_{pj}^2 + \sigma_{ij,pij,e}^2$$

Description: Each person takes a different set of items and all persons' item responses are evaluated by all of the judges. Items are nested within persons, and items and persons are crossed with judges.

Part 3: Decomposition of Score Effects and Variance for the $i{:}(p \times j)$ Design

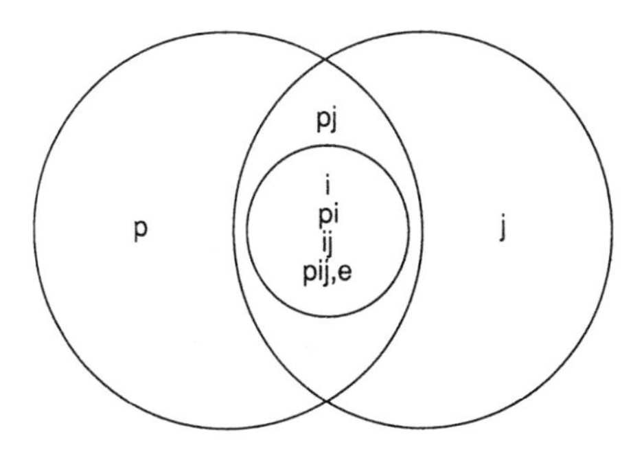

Figure 4.6. Sources of Variability

$$
\begin{aligned}
X_{pij} = \quad & \mu \\
& + (\mu_p - \mu) \\
& + (\mu_j - \mu) \\
& + (\mu_{pj} - \mu_p - \mu_j + \mu) \\
& + (X_{pij} - \mu_{pj}) \\
\sigma^2(X_{pij}) = \; & \sigma_p^2 + \sigma_j^2 + \sigma_{pj}^2 + \sigma_{i,pi,ij,pij,e}^2
\end{aligned}
$$

Description: Each person is evaluated by all judges, but the subset of the 20 items is different for each combination of persons and judges. Persons are crossed with judges. Items are nested within the combination of persons and judges.

Part 4: Decomposition of Score Effects and Variance for the $(i \times j){:}p$ Design

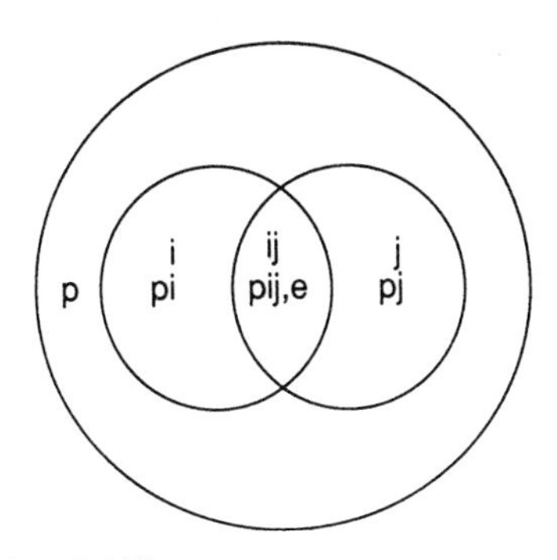

Figure 4.7. Sources of Variability

$$
\begin{aligned}
X_{pij} = \quad & \mu \\
& + (\mu_p - \mu) \\
& + (\mu_{pi} - \mu_p) \\
& + (\mu_{pj} - \mu_p) \\
& + (X_{pij} - \mu_{pi} - \mu_{pj} + \mu_p)
\end{aligned}
$$

$$\sigma^2 (X_{pij}) = \sigma_p^2 + \sigma_{i,pi}^2 + \sigma_{j,pj}^2 + \sigma_{ij,pij,e}^2$$

Description: All items are evaluated by all judges, but each person gets a different item–judge combination. For example, one person may take items 1 through 5, which are evaluated by judges 3 and 4; another person may take items 6 through 10, which are evaluated by judges 1 and 2; another person may take items 11 through 15, which are evaluated by judges 5 and 6; another person may take items 16 through 20, evaluated by judges 7 and 8; and so on. Items and judges are crossed; both are nested within persons.

Part 5: Decomposition of Score Effects and Variance for the *i*:*j*:*p* Design

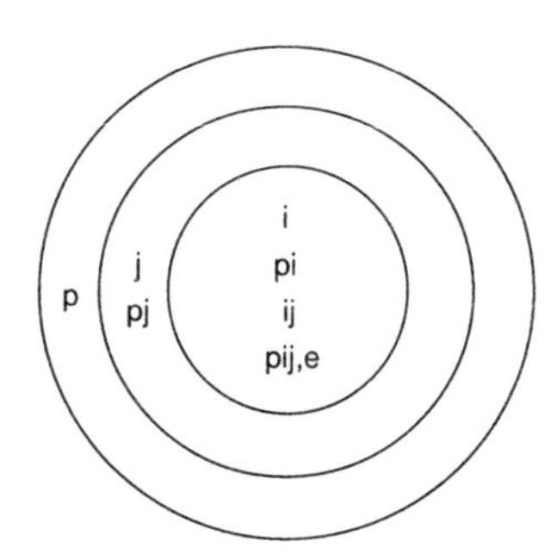

Figure 4.8. Sources of Variability

$$
\begin{aligned}
X_{pij} = \quad & \mu \\
& + (\mu_p - \mu) \\
& + (\mu_{pj} - \mu_p) \\
& + (X_{pij} - \mu_{pj})
\end{aligned}
$$

$$\sigma^2 (X_{pij}) = \sigma_p^2 + \sigma_{j,pj}^2 + \sigma_{i,pi,ij,pij,e}^2$$

Description: Each person is evaluated by a different set of judges, and each judge evaluates a different set of items. Items are nested within judges and judges are nested within persons. It follows that items are nested also within persons; that is, each person is administered a different set of items.

Appendix 4.3

Expected Mean Squares Equations for Nested, Two-Facet Random-Effects Designs

***p* × (*i*:*j*) Design** [a]

$$EMS_p = n_i n_j \sigma^2_p + n_i \sigma^2_{pj} + \sigma^2_{pi,pij,e}$$
$$EMS_j = n_p n_i \sigma^2_j + n_p \sigma^2_{i,ij} + \sigma^2_{pi,pij,e}$$
$$EMS_{pj} = n_i \sigma^2_{pj} + \sigma^2_{pi,pij,e}$$
$$EMS_{i,ij} = n_p \sigma^2_{i,ij} + \sigma^2_{pi,pij,e}$$
$$EMS_{pi,pij,e} = \sigma^2_{pi,pij,e}$$

(*i*:*p*) × *j* Design [b]

$$EMS_p = n_i n_j \sigma^2_p + n_j \sigma^2_{i,pi} + \sigma^2_{ij,pij,e}$$
$$EMS_j = n_p n_i \sigma^2_j + n_i \sigma^2_{pj} + \sigma^2_{ij,pij,e}$$
$$EMS_{pj} = n_i \sigma^2_{pj} + \sigma^2_{ij,pij,e}$$
$$EMS_{i,pi} = n_j \sigma^2_{i,pi} + \sigma^2_{ij,pij,e}$$
$$EMS_{ij,pij,e} = \sigma^2_{ij,pij,e}$$

***i*: (p × *j*) Design** [c]

$$EMS_p = n_i n_j \sigma^2_p + n_i \sigma^2_{pj} + \sigma^2_{i,pi,ij,pij,e}$$
$$EMS_j = n_p n_i \sigma^2_j + n_i \sigma^2_{pj} + \sigma^2_{i,pi,ij,pij,e}$$
$$EMS_{pj} = n_i \sigma^2_{pj} + \sigma^2_{i,pi,ij,pij,e}$$
$$EMS_{i,pi,ij,pij,e} = \sigma^2_{i,pi,ij,pij,e}$$

(*i* × *j*):*p* Design

$$EMS_p = n_i n_j \sigma^2_p + n_j \sigma^2_{i,pi} + n_i \sigma^2_{j,pj} + \sigma^2_{ij,pij,e}$$
$$EMS_{i,pi} = n_j \sigma^2_{i,pi} + \sigma^2_{ij,pij,e}$$
$$EMS_{j,pj} = n_i \sigma^2_{j,pj} + \sigma^2_{ij,pij,e}$$
$$EMS_{ij,pij,e} = \sigma^2_{ij,pij,e}$$

***i*:*j*:*p* Design** [d]

$$EMS_p = n_i n_j \sigma^2_p + n_i \sigma^2_{j,pj} + \sigma^2_{i,pi,ij,pij,e}$$
$$EMS_{j,pj} = n_i \sigma^2_{j,pj} + \sigma^2_{i,pi,ij,pij,e}$$
$$EMS_{i,pi,ij,pij,e} = \sigma^2_{i,pi,ij,pij,e}$$

---

a. For $p \times (j{:}i)$ design, switch the i and j subscripts.
b. For $(j{:}p) \times i$ design, switch the i and j subscripts.
c. For $j{:}(p \times i)$ design, switch the i and j subscripts.
d. For $j{:}i{:}p$ design, switch the i and j subscripts.

Exercises

1. In a study of classroom interaction, Erlich and Borich (1979) observed 17 second- and third-grade classrooms each on five occasions. Because "teachers were observed at different times of day, on different days and teaching what may be considered different lessons," Erlich and Borich (1979, p. 12) treated occasions as nested within teachers. From the results of the ANOVA for the behavior "teacher asks a new question following correct response to the teacher's previous questions," estimate and interpret the variance components. See Table 4.3.

TABLE 4.3

Source of Variation	*Sum of Squares*	*df*	*Mean Square*
Teachers (p)	1299.98	16	81.25
Occasions:Teachers (o:p)	90.92	4	22.73

Source: Data are from "Occurrence and Generalizability of scores on a classroom interaction instrument," by O. Erlich and G. Borich, *Journal of Educational Measurement*, 1979, *16*.

2. In a study of Marine Corps infantry riflemen, 43 riflemen were observed by two raters on three occasions as they assembled communications equipment. A different pair of raters evaluated the riflemen's performance on each occasion, but all riflemen performed the task on every occasion. What was the design of the study?
3. The results of the ANOVA in the G study of Marine Corps infantry riflemen assembling communications equipment are presented in Table 4.4.

TABLE 4.4

Source of Variation	*Sum of Squares*	*df*	*Mean Square*
Persons (p)	4.0908	42	0.0974
Occasions (o)	1.3606	2	0.6803
Raters:Occasions (r:o)	0.2139	3	0.0713
po	1.8816	84	0.0224
pr,pro,e	1.9558	127	0.0154

a. List the expected mean squares for this design.

b. Calculate and interpret the estimated variance components.

4. Teachers (p) were evaluated on multiple items (i) of teaching performance by multiple raters (r). The object of measurement is teachers; the facets are items and raters. Describe the possible two-facet designs that could be used, and list the variance components associated with each.
5. What are the disadvantages of the $(r{:}p) \times i$ design in Exercise 4 above for assessing separate sources of variance?

Answers to Exercises

1. $\hat{\sigma}_p^2 = 11.70$ (34%), $\hat{\sigma}_{o,po,e}^2 = 22.73$ (66%). The substantial universe-score variation shows that teachers differed in their behavior. The very large residual component (o,po,e) is due to large differences in teacher behavior across occasions, unmeasured systematic or unsystematic variation, or both.
2. $p \times (r{:}o)$. Persons were crossed with raters and occasions, and raters were nested within occasions (see Figure 4.1 (a) in the text).
3. a.

$$\begin{aligned}
\text{EMS}\,(p) &= \sigma_{pr,pro,e}^2 + 2\sigma_{po}^2 + 6\sigma_p^2 \\
\text{EMS}\,(o) &= \sigma_{pr,pro,e}^2 + 2\sigma_{po}^2 + 43\sigma_{r,ro}^2 + 86\sigma_o^2 \\
\text{EMS}\,(r,ro) &= \sigma_{pr,pro,e}^2 + 43\sigma_{r,ro}^2 \\
\text{EMS}\,(po) &= \sigma_{pr,pro,e}^2 + 2\sigma_{po}^2 \\
\text{EMS}\,(pr,pro,e) &= \sigma_{pr,pro,e}^2
\end{aligned}$$

b.

	p	o	r,ro	po	pr,pro,e
$\hat{\sigma}^2 =$	0.0125	0.0070	0.0013	0.0035	0.0154
% =	32	18	3	9	39

Substantial differences were found between riflemen in their performance (large p effect). Raters differed little in the scales they used to evaluate riflemen, and raters used the same scales on different occasions (small r,ro effect). Average riflemen performance differed substantially across occasions (large o effect), and the relative standing of riflemen differed somewhat across occasions (nontrivial po effect). The large residual effect (pr,pro,e) indicates substantial changes in the relative standing of riflemen over raters (pr) or over rater–occasion combinations (pro), and/or large unmeasured variation (e).

4. (1) $p \times r \times i$ design: Each teacher is evaluated by every rater on every item.

$$\sigma_p^2,\ \sigma_r^2,\ \sigma_i^2,\ \sigma_{pr}^2,\ \sigma_{pi}^2,\ \sigma_{ri}^2,\ \sigma_{pri,e}^2$$

(2) $(r{:}p) \times i$ design: Each teacher is evaluated by a different set of raters. But every rater uses, and every teacher is evaluated on, all items.

$$\sigma_p^2,\ \sigma_i^2,\ \sigma_{pi}^2,\ \sigma_{r,pr}^2,\ \sigma_{ri,pri,e}^2$$

(3) $(i{:}p) \times r$ design: Same as above but switch r and i subscripts.

(4) $p \times (i{:}r)$ design: Each teacher is evaluated by all raters and on all items, but each rater uses a different set of items.

$$\sigma_p^2,\ \sigma_r^2,\ \sigma_{pr}^2,\ \sigma_{i,ri}^2,\ \sigma_{pi,pri,e}^2$$

(5) $p \times (r{:}i)$ design: Same as above but switch r and i subscripts.

(6) $i{:}r{:}p$ design: Each teacher is evaluated by a different set of raters, and every rater uses a different set of items.

$$\sigma_p^2,\ \sigma_{r,pr}^2,\ \sigma_{i,ri,pi,pri,e}^2$$

(7) $r{:}i{:}p$ design: Same as above but switch r and i subscripts.

5. The confounding of the rater (r) effect and the teacher-by-rater (pr) effect makes it impossible to determine whether some raters are more lenient than others in their judging of teacher performance (r effect) or whether the relative standing of teachers differs across raters (pr effect). The confounding of the ri and pri,e effects produces similar ambiguity.

5

Generalizability Studies with Fixed Facets

Until now we have assumed that all sources of variation were random, and the intent was to generalize to all items, occasions, and observers in a larger universe, not just those used in the generalizability study. Sometimes, however, it is not appropriate or not possible to generalize beyond the conditions of a facet. One reason is that for substantive, cost, or logistical reasons, the investigator specifically selects for inclusion certain conditions of a facet. A second reason is that the facet has only a few conditions and all are included in a G study. In both cases, the facet is considered fixed, not random. This chapter develops the measurement model for G studies with fixed facets. It explains when facets should be considered fixed, and how to estimate variance components and interpret them in G studies with fixed facets.

Definition

A fixed facet in G theory is analogous to a fixed factor in ANOVA. Just as the levels of a fixed factor in ANOVA exhaust all levels in the universe to which an inference will be drawn, the conditions of a fixed facet in a G study exhaust all possible conditions of interest in the universe. A fixed facet may arise in two ways: (a) The decision maker purposively selects certain conditions from the universe for the G study and is not interested in generalizing beyond them (or it is not reasonable to do so), or (b) the entire universe of conditions is small and all conditions are included in the measurement design.

As an example of (a) "purposeful selection," Erlich and Shavelson (1978) conducted a generalizability study of observations of teacher behavior in two subject matters. Fifth-grade teachers were videotaped during a reading lesson and during a math lesson. Their behavior in "asking students to try again" was coded by multiple raters. The facets of this study were raters and subject matter of the lesson. Rater was considered to be a random facet. Even if the raters used in the G study were not selected randomly from the universe of raters, they could be considered exchangeable with other raters in the universe.

The same could not be said for subject matter. Reading and mathematics lessons were selected purposively because they represented important areas of the elementary school curriculum to which all teachers devote time. They were not selected at random from the universe of subject matters that teachers teach, and they could not be considered exchangeable with other subject matters (e.g., art, science, social studies). There is no reason to believe that the behavior of teachers when teaching reading and mathematics generalizes to their behavior when teaching other subject matters. Consequently, Erlich and Shavelson considered subject matter to be a *fixed* facet.

As an example of (b) "entire universe sampled," teachers in the Erlich and Shavelson (1978) study could have been videotaped while teaching all possible subject matters. The list would have been fairly small (probably less than 10). In this case, the subject matters sampled in the generalizability study would have exhausted the universe of subject matter conditions. By definition, the decision maker could not generalize beyond the conditions sampled in the G study because the universe had been exhausted. Subject matter would be considered a fixed facet in this situation, too.

It is important to note that G theory is essentially a random facet measurement theory; that is, it considers only designs in which at least one facet of error is random. All facets may be random (a random model), or some facets may be random and some fixed (a mixed model). All facets cannot be fixed (a fixed model). All models in generalizability theory that have a fixed facet, then, are mixed models.

Analysis of Designs with Fixed Facets

Statistically, G theory treats a fixed facet by averaging over the conditions of the facet. For example, in the Erlich and Shavelson (1978) study with subject matter as a fixed facet, the score for a teacher averaged over reading and math may provide the best score for a teacher because it represents the teacher's universe score over all the conditions of the fixed facet (here, reading and math). This information may be useful to elementary school principals, for example, who are interested in a teacher's behavior in the major subject matters taught.

Other users, however, may believe that instruction is typically different in reading and mathematics and, consequently, feel that it does not make sense to average over them. If it does not make conceptual sense to average over the conditions of a fixed facet, or if conclusions about such averages are of little interest, separate G studies should be conducted within each condition of the fixed facet. This separation permits different conclusions to be reached about the measurement of teaching in reading and in mathematics.

We propose that the decision to choose between approaches be made primarily on conceptual grounds. The decision maker should decide what kind of information will be needed and should analyze the data accordingly. If the investigator chooses to average over conditions of the fixed facet, however, we recommend that he or she also examine the variance components associated with the fixed facet. Large variation for the fixed facet indicates that the conditions of the fixed facet differ in substantial ways. It may make sense, then, to analyze each condition separately, too.

Averaging over Conditions of the Fixed Facet. The procedure for averaging over the fixed facet has three steps.[1]

Step 1. Run an analysis of variance treating all sources of variance as random (even those that are fixed) to estimate the variance components from that fully random analysis. For example, in the $p \times r \times s$ design with persons (p) crossed with raters (r) and subject matters (s), persons and raters are random and subject matter is fixed. For this step, however, treat all sources of variance, including subject matter, as random. The variance components to be estimated in the fully random $p \times r \times s$ design are σ^2_p, σ^2_r, σ^2_s, σ^2_{pr}, σ^2_{ps}, σ^2_{rs}, and $\sigma^2_{prs,e}$.

Step 2. Identify the random portion of the mixed design and the associated variance components to be calculated. In the $p \times r \times s$ design with persons (p) and raters (r) random and subject matter (s) fixed, the random portion of the design is persons (p) crossed with raters (r). The variance components to be calculated for the random portion of the design, then, correspond to persons (p), raters (r), and the interaction between persons and raters, plus remaining error (pr,e). We will call these variance components σ^2_{p*}, σ^2_{r*}, and $\sigma^2_{pr,e*}$ to distinguish them from the variance components computed in the fully random design in Step 1.

Step 3. Calculate the variance components for the random portion of the mixed design identified in Step 2. Each variance component is the variance component for that source from the fully random design in Step 1 *plus* $1 / n_s$ times the variance component corresponding to the interaction between that source of variance and the fixed facet (where n_s is the number of conditions of the fixed facet s). The variance components to be estimated here are the following:

$$\sigma^2_{p*} = \sigma^2_p + \frac{1}{n_s}\sigma^2_{ps}$$

$$\sigma^2_{r*} = \sigma^2_r + \frac{1}{n_s}\sigma^2_{rs}$$

$$\sigma^2_{pr,e*} = \sigma^2_{pr} + \frac{1}{n_s}\sigma^2_{prs,e}$$

Note that the right-hand side of each equation includes interactions with *only* the fixed facet, *not* any interactions that include a random facet. For example, in the calculation of σ^2_{p*}, the term $\sigma^2_{prs,e}$ is not included because it involves facet r, a random facet. Appendix 5.1 summarizes the steps for other two-facet designs with a fixed facet.

Analyzing Each Condition of the Fixed Facet Separately. If a decision maker decides that it does not make sense conceptually to average over conditions of the fixed facet, or if the estimated variance components associated with the fixed facets ($\hat{\sigma}^2_s$, $\hat{\sigma}^2_{ps}$, $\hat{\sigma}^2_{rs}$, and $\hat{\sigma}^2_{prs,e}$ in Step 1 above) are large, we recommend that the G study analyze and report results for each condition of the fixed facet separately. When the fixed facet is subject matter, for example, the data would be analyzed separately for reading and mathematics. The analysis for each subject matter will have a $p \times r$, random-effects design.

Numerical Example: Crossed Design

To illustrate the analysis of data from a crossed, two-facet design with a fixed facet, we use a hypothetical data set in which eight teachers are each observed by three raters for two subject matters: reading and mathematics. Teachers (p) and raters (r) are considered random. Subject matter (s) is considered fixed because (a) these are subject matters specifically selected to represent standard curriculum areas, and (b) it is probably not reasonable to generalize from teachers' behavior when

TABLE 5.1 Data for Two-Facet Crossed $p \times r \times s$ Study of Teacher Behavior

	Subject Matter:	*Mathematics*			*Reading*		
Teacher	*Rater:*	*1*	*2*	*3*	*1*	*2*	*3*
1		4	4	4	5	5	6
2		6	7	6	7	9	5
3		8	7	7	4	3	2
4		6	8	7	9	11	7
5		2	1	1	5	5	3
6		5	4	4	7	6	5
7		4	5	6	6	8	9
8		7	7	6	5	9	9

teaching mathematics and reading lessons to their behavior when teaching other subject matters. For purposes of illustration, we carry out two procedures here: averaging over subject matters, and analyzing each subject matter separately.

Averaging over Subject Matters. The first step is to perform the analysis of variance for the $p \times r \times s$ design, treating all facets as random.

The data are given in Table 5.1, and the results of the fully random analysis are given in Table 5.2. The main effect for subject matter is small ($\sigma_s^2 = 0.1577$; only 3% of the total variance). Teacher behavior was similar in reading and mathematics, averaging over teachers and raters. The interaction between teachers and subject matters, however, is very large ($\sigma_{ps}^2 = 2.3423$; 46% of the total variance). This result shows

TABLE 5.2 Two-Facet Crossed $p \times r \times s$ Design Treating All Facets as Random (Step 1 in Analysis of Design with Fixed Facet)

Source of Variation	*df*	*Mean Squares*	*Variance Component*	*Estimated Variance Component*	*Percentage of Total Variance*
Teachers (p)	7	15.9524	σ_p^2	1.1726	23
Raters (r)	2	2.4375	σ_r^2	0.0149	0
Subject matters (s)	1	12.0000	σ_s^2	0.1577	3
pr	14	1.8899	σ_{pr}^2	0.5060	10
ps	7	7.9048	σ_{ps}^2	2.3423	46
rs	2	1.1875	σ_{rs}^2	0.0387	1
prs,e	14	0.8780	$\sigma_{prs,e}^2$	0.8780	17

that the relative standing of teachers differed substantially from reading to mathematics. Because the behavior of teachers differed substantially in the two subject matters, it may be reasonable to analyze each subject matter separately, as well as averaging over them. (The results of the separate analyses are presented in the next section.)

The second step is to identify the random portion of the mixed design. The random portion of this design has teachers (p) crossed with raters (r) or a $p \times r$ design. The third step is to calculate the variance components associated with the $p \times r$ design:

$$\sigma^2_{p*} = \sigma^2_p + \frac{1}{n_s}\sigma^2_{ps} \quad [5.1]$$

$$\sigma^2_{r*} = \sigma^2_r = \frac{1}{n_s}\sigma^2_{rs} \quad [5.2]$$

$$\sigma^2_{pr,e*} = \sigma^2_{pr} + \frac{1}{n_s}\sigma^2_{prs,e} \quad [5.3]$$

The variance components on the left-hand side of Equations 5.1, 5.2, and 5.3 are the variance components for the mixed design; the estimated variance components on the right-hand side of Equations 5.1, 5.2, and 5.3 are those produced by the fully random $p \times r \times s$ analysis; and n_s is the number of subject matters comprising the subject matter facet (here, 2).

The resulting estimated variance components for the mixed design appear in Table 5.3. Universe-score variance accounts for the lion's share of the total variation in the measurement. The residual component ($\hat{\sigma}^2_{pr,e*} = 0.9450$, 28% of the total variance), however, is also large, which suggests that multiple raters will be necessary to obtain generalizable scores averaged over reading and mathematics. This issue is discussed in the next chapter.

Analyzing Each Subject Matter Separately. The separate generalizability analyses for mathematics and reading appear in Table 5.4. The analyses of mathematics and reading produced somewhat different results. Most of the variation in mathematics is due to universe-score variation, whereas in reading substantial variation is due to both universe-score and residual variation. This difference will lead to different

TABLE 5.3 Analysis of the $p \times r \times s$ Design with s Fixed

Random Design			*Mixed Design with s Fixed* [a]			
Source of Variation	*Variance Component*	*Estimated Variance Component*	*Source of Variation*	*Variance Component*	*Estimated Variance Component*	*Percentage of Total Variance*
Teachers (p)	σ_p^2	1.1726	Teachers (p^*)	$\sigma_{p^*}^2$	2.3438	71
Raters (r)	σ_r^2	0.0149	Raters (r^*)	$\sigma_{r^*}^2$	0.0343	1
Subject matters (s)	σ_s^2	0.1577				
pr	σ_{pr}^2	0.5060	pr,e^*	σ_{pr,e^*}^2	0.9450	28
ps	σ_{ps}^2	2.3423				
rs	σ_{rs}^2	0.0387				
prs,e	$\sigma_{prs,e}^2$	0.8780				

a. Averaging over levels of the fixed facet.

TABLE 5.4 Separate Analyses of Teacher Behavior by Subject Matter

		Mathematics		*Reading*	
Source of Variation	*Variance Component*	$\hat{\sigma}^2$	*Percentage*	$\hat{\sigma}^2$	*Percentage* [a]
Teachers (p)	σ_p^2	3.8869	88	3.1429	57
Raters (r)	σ_r^2	0[b]	0	0.1548	3
pr,e	$\sigma_{pr,e}^2$	0.5060	12	2.2619	41

a. Percentage of total variance; percentages do not sum exactly to 100 due to rounding.
b. Variance component of –0.0476 set equal to zero.

decisions about the optimal number of raters to use in the two subject matters, as will be discussed in the next chapter.

Numerical Example: Partially Nested Design with a Fixed Facet

G theory can be applied to designs with both nested facets and fixed facets. Consider a study of the personality construct self-concept (e.g, Shavelson, Hubner, & Stanton, 1976). Self-concept refers to an individual's perception and evaluation of himself or herself. Marsh's (in press) Self-Description Questionnaire (SDQ) assesses numerous self-concept dimensions in açademic and nonacademic areas. Three of the dimensions are general self-concept, academic self-concept, and mathematics self-concept. The general, academic, and mathematics scales of the SDQ have 16, 12, and 10 items each, respectively. On each item, individuals rate themselves on a 6-point scale ranging from 1 (false) to 6 (true). For example, an item from the math self-concept scale is

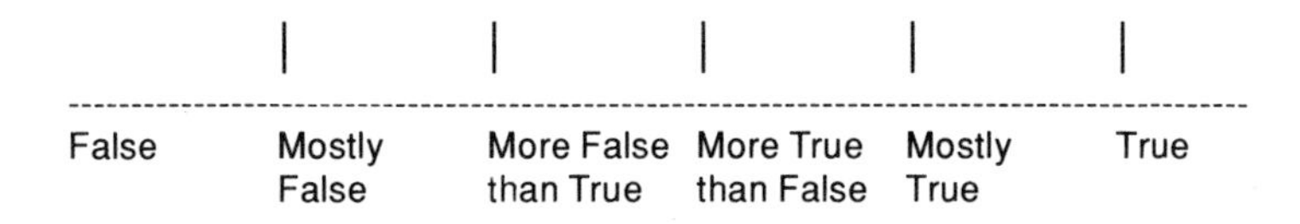

In the G study reported here, the three scales of the SDQ were administered to 140 seventh-grade students. In this design, the items (i) are nested within scale (s) because there are multiple items per scale and the items differ from scale to scale. Both items and scales are crossed with persons (p) because each person responded to all items in all scales. The notation for this design is $p \times (i{:}s)$.

These scales were selected to represent three different levels of self-concept: The general scale contains items about one's overall perception of oneself; the academic scale contains items about one's perceived abilities in school subjects; and the mathematics scale refers to one's perceived competence in mathematics. No intention was made to generalize beyond these areas of self-concept to others (e.g., self-concept of physical ability or appearance, or self-concept in other subject matter areas), and it would not be reasonable to do so. Hence, the scale facet is appropriately considered a fixed facet, not a random one.

Both averaging over scales and analyzing each scale separately might make conceptual sense here. The dimensions of self-concept may be distinct and can be analyzed separately. Or they may be components of a single undifferentiated self-concept variable, well represented by the average of the scales. For purposes of illustration, then, both procedures are carried out here.

Averaging over Scales. The first step in the G study is to obtain ANOVA estimates of variance components treating all facets, including scale, as random. Because the scales have different numbers of items, the design is *unbalanced.* For the fully random analysis with all three scales, we selected 10 items at random from each scale to create a *balanced* design. The reason for doing so is that most computer programs (e.g., BMDP 3V and MIVQUEO in the VARCOMP procedure of SAS) do not have sufficient storage capacities to analyze typical unbalanced designs (Brennan, Jarjoura, & Deaton, 1980). With unbalanced data, the algebraic formulas for estimating variance components are complicated (Searle, 1971), which makes estimation of the variance components with unbalanced data computationally complex.

Randomly deleting levels of a facet to create a balanced design has little effect on the estimated variance components (see Shavelson, Webb, & Rowley, 1989) and allows a wide variety of computer programs to be used. When randomly deleting data, however, we recommend comparing the results of several random deletions to make sure that any particular selection is not unusual. For the SDQ data set, we estimated variance components for several randomly selected sets of 10 items per scale. The estimated variance components were very similar for all sets of randomly selected items.

The results of the first step in the analysis, treating all facets as random, appear in Table 5.5 . The main effect of scale is nonnegligible (σ_s^2 = 0.2040, 7% of the total variance). It shows that some scales yielded somewhat higher average self-concept scores than others. The more substantial effect for the interaction between persons and scales (σ_{ps}^2 = 0.3121; 11% of the total variance) shows that the relative standing of persons differed from scale to scale. Based on this information, the decision maker may decide to analyze each scale separately.

TABLE 5.5 Two-Facet, Partially Nested $p \times (i{:}s)$ Design Treating All Facets as Random (Step 1 in Analysis of Design with Fixed Facet)

Source of Variation	*df*	*Mean Squares*	*Variance Component*	*Estimated Variance Component*	*Percentage of Total Variance*
Persons (p)	139	19.18	σ_p^2	0.4858	18
Scales (s)	2	327.61	σ_s^2	0.2040	7
Items ($i{:}s$)	27	38.83	$\sigma_{i:s}^2$	0.2668	10
ps	278	4.61	σ_{ps}^2	0.3121	11
pi:s,e	3753	1.49	$\sigma_{pi:s,e}^2$	1.4870	54

It should be noted that, in contrast to the notation used in the nested designs in chapter 4, Table 5.5 uses the notation $\sigma_{i:s}^2$ instead of $\sigma_{i,is}^2$, and $\sigma_{pi:s,e}^2$ instead of $\sigma_{pi,pis,e}^2$. Because items are nested within scales in the universe, it is not meaningful to think about the effect *i:s* as the confounding of separate effects *i* and *is*. No separate *i* effect exists because an item score cannot be interpreted independently of the scale it is in. And no separate $i \times s$ interaction effect exists because one cannot think of items having different relative standings across scales—items belong to only one scale.

The nested designs illustrated in chapter 4 [$i{:}p$ and $(o{:}p) \times r$], in contrast, assumed crossed $p \times i$ and $p \times r \times o$ universes. In a crossed $p \times i$ universe, the *i* and *pi* effects have distinct interpretations, so it makes sense to talk about the confounding of the separate *i* and *pi* effects in the nested *i:p* design. Similarly, it makes sense to talk about confounding of the separate *o* and *po* effects and the confounding of the separate *ro* and *pro,e* effects in the partially nested $(o{:}p) \times r$ design.

The second step in the G study is to identify the random portion of the mixed design (with scale considered fixed) and the associated variance components. The random portion of the design has persons crossed with items—$p \times i$.

The third step is to calculate the variance components for the $p \times i$ design:

TABLE 5.6 Analysis of the $p \times (i{:}s)$ Design with s Fixed

Random $p \times (i{:}s)$ Design			*Mixed $p \times (i{:}s)$ Design with s Fixed* [a]		
Source of Variation	*Variance Component*	*Estimated Variance Component*	*Variance Component*	*Estimated Variance Component*	*Percentage of Total Variance*
Persons (p)	σ^2_p	0.4858	σ^2_{p*}	0.5898	25
Scales (s)	σ^2_s	0.2040			
Items ($i{:}s$)	$\sigma^2_{i:s}$	0.2668	σ^2_{i*}	0.2668	11
ps	σ^2_{ps}	0.3121			
$pi{:}s,e$	$\sigma^2_{pi:s,e}$	1.4870	$\sigma^2_{pi,e*}$	1.4870	64

a. Averaging over levels of the fixed facet.

$$\sigma^2_{p*} = \sigma^2_p + \frac{1}{n_s}\sigma^2_{ps} \qquad [5.4]$$

$$\sigma^2_{i*} = \sigma^2_{i:s} \qquad [5.5]$$

$$\sigma^2_{pi,e*} = \sigma^2_{pi:s,e} \qquad [5.6]$$

The variance components on the left-hand side of Equations 5.4, 5.5, and 5.6 are the variance components for the mixed design; the estimated variance components on the right-hand side of Equations 5.4, 5.5, and 5.6 are those produced by the fully random $p \times (i{:}s)$ analysis; and n_s is the number of scales comprising the scale facet (here, 3).

The results from all of these steps are presented in Table 5.6. Differences between items were substantial ($\hat{\sigma}^2_{i*} = 0.2668$, 11% of the total variance). The large residual component (64% of the total variance) indicates large differences in the relative standing of persons on different items, large unmeasured variation, or both.

Analyzing Each Scale Separately. A separate generalizability analysis with a $p \times i$ design was carried out for each scale (Table 5.7). The three separate analyses allowed us to use all of the items in the original scales (16, 12, 10), not the 10 items randomly selected to create a

TABLE 5.7 Separate Analyses of Three Scales of the SDQ

		General		*Academic*		*Mathematics*	
Source of Variation	*Variance Component*	$\hat{\sigma}^2$	*Percentage*[a]	$\hat{\sigma}^2$	*Percentage*[a]	$\hat{\sigma}^2$	*Percentage*[a]
Persons (p)	σ_p^2	0.6100	30	0.8054	31	0.9761	35
Items (i)	σ_i^2	0.1228	6	0.3585	14	0.2363	8
pi,e	$\sigma_{pi,e}^2$	1.3189	64	1.4093	55	1.6071	57

a. Percentage of total variance.

balanced $p \times (i{:}s)$ design for Table 5.5. The analyses using the 10 randomly selected items were also carried out; they yielded similar results and so are not presented here.

As can be seen in Table 5.7, the separate analyses of the three self-concept scales yielded fairly similar patterns of estimated variance components. The variance components for persons accounted for similar proportions of variance across the three scales (between 30% and 35% of the total variance for each scale). Each scale had substantial differences among persons in their ratings of their self-concept. The variance components for items accounted for a smaller share of the variance (ranging from 6% to 14% across scales). The general self-concept scale showed the least variation across items, while the academic self-concept scale showed the greatest variation across items. The residual component ($\sigma_{pi,e}^2$) accounted for most of the variation in each scale (ranging from 55% to 64%): (a) The relative standing of persons changed from item to item (a large $p \times i$ interaction), (b) the unmeasured effects in this study (systematic or unsystematic) were substantial, or (c) both.

Summary

Until this chapter, an assumption has been made that either the conditions of a facet have been sampled randomly or the observed conditions are exchangeable with any other of an indefinitely large number of conditions. In some facets, however, the conditions are small in number (e.g., subtests of a general aptitude test) or the investigator is (a) substantively interested in performance only under a certain, small number of the conditions (e.g., occasions within the first week of school); or (b) cost or logistical concerns prohibit sampling from a large number of conditions of a particular facet. In this case we speak of a *fixed* facet, a facet with a small number of conditions, all of which are observed in the G study. For example, in studies of elementary school teaching, the focus is on teachers' behavior in certain subject matters. Several raters might observe a set of elementary teachers teaching, say, math lessons and reading lessons. In this case, the rater facet may be considered random (either sampled randomly or exchangeable with other raters). The subject matter facet, however, is properly considered fixed. Math and reading exhaust the universe to which a decision maker might wish to generalize.

G theory treats fixed facets specially. In general, G theory averages over the conditions of the fixed facet; for example, the average of a teacher's behavior in math and in reading. Sometimes, however, it does not make sense to average over the conditions of a fixed facet, or interest may not lie in the average. For example, if the decision maker believes that teaching math is different from teaching reading, then the average would not make conceptual sense. In this case a separate G study should be conducted for each condition of the fixed facet: one with observations made during math lessons and one with observations made during reading lessons.

These two procedures can be combined into a general strategy. If the decision maker chooses to average over levels of the fixed facet, we recommend that variation due to the fixed facet be examined. If it is large, the decision maker may decide also to analyze separately each condition of the fixed facet.

Appendix 5.1

Calculating Variance Components in Two-Facet Designs with a Fixed Facet

Parts 1 to 4 of this appendix summarize the steps in calculating the estimated variance components for four different two-facet designs with a fixed facet. These are the only possible two-facet designs with a fixed facet j because a fixed facet cannot be nested within a random source of variance (whether the object of measurement or a random facet of error). Consider the impossible case of fixed facet j (say, subject matter) nested within random facet i (say, raters). In this case, each rater (i) would rate a different set of subject matters (j). The sets of subject matters are not considered exchangeable with each other because this facet is fixed. Consequently, the pairing of sets of subject matters with raters are not considered to be exchangeable. But raters, being random, *are* considered to be exchangeable with each other, leading to a contradiction.

Part 1: Estimated Variance Components for $p \times i \times j$ Design with j Fixed

1. Variance components in fully random $p \times i \times j$ design:

$$\sigma_p^2, \sigma_i^2, \sigma_j^2, \sigma_{pi}^2, \sigma_{pj}^2, \sigma_{ij}^2, \sigma_{pij,e}^2$$

2. Random part of $p \times i \times j$ design with j fixed: $p \times i$
3. Variance components to be estimated:

$$\sigma_{p*}^2 = \sigma_p^2 + \frac{1}{n_j} \sigma_{pj}^2$$

$$\sigma_{i*}^2 = \sigma_i^2 + \frac{1}{n_j} \sigma_{ij}^2$$

$$\sigma_{pi,e*}^2 = \sigma_{pi}^2 + \frac{1}{n_j} \sigma_{pij,e}^2$$

Part 2: Variance Components for $p \times (i{:}j)$ Design with j Fixed

1. Variance components in fully random $p \times (i{:}j)$ design:

$$\sigma_p^2, \sigma_j^2, \sigma_{pj}^2, \sigma_{i,ij}^2, \sigma_{pi,pij,e}^2$$

2. Random part of $p \times (i{:}j)$ design with j fixed: $p \times i$
3. Variance components to be estimated:

$$\sigma_{p*}^2 = \sigma_p^2 + \frac{1}{n_j}\sigma_{pj}^2$$

$$\sigma_{i*}^2 = \sigma_{i,ij}^2$$

$$\sigma_{pi,e*}^2 = \sigma_{pi,pij,e}^2$$

Part 3: Estimated Variance Components for a $(i{:}p) \times j$ Design with j Fixed

1. Variance components in fully random $(i{:}p) \times j$ design:

$$\sigma_p^2, \sigma_j^2, \sigma_{pj}^2, \sigma_{i,pi}^2, \sigma_{ij,pij,e}^2$$

2. Random part of $(i{:}p) \times j$ design with j fixed: $i{:}p$
3. Variance components to be estimated:

$$\sigma_{p*}^2 = \sigma_p^2 + \frac{1}{n_j}\sigma_{pj}^2$$

$$\sigma_{i,pi,e*}^2 = \sigma_{i,pi}^2 + \frac{1}{n_j}\sigma_{ij,pij,e}^2$$

Part 4: Estimated Variance Components for a $i{:}(p \times j)$ Design with j Fixed

1. Variance components in fully random $i{:}(p \times j)$ design:

$$\sigma_p^2, \sigma_j^2, \sigma_{pj}^2, \sigma_{i,pi,ij,pij,e}^2$$

2. Random part of $i{:}(p \times j)$ design with j fixed: $i{:}p$
3. Variance components to be estimated:

$$\sigma_{p*}^2 = \sigma_p^2 + \frac{1}{n_j}\sigma_{pj}^2$$

$$\sigma_{i,pi,e*}^2 = \sigma_{i,pi,ij,pij,e}^2$$

Exercises

1. In a study of teacher effectiveness, Marzano (1973) audiotaped teachers during five 1-week segments over a 6-month period. During each 1-week segment, teachers were observed while teaching lessons from books A, B, and C of the *Distar Language 1* program. The design was fully crossed: teachers (p) by occasions (o) by books (b).

   For each scenario below, decide whether books should be considered random or fixed.

   a. Books A, B, C were purposively selected from all Distar books because they cover the most basic skills.

   b. Books A, B, C were selected because they are the books most often used by teachers.

   c. Books A, B, C are selected because the schools in the study were already using them.

   d. Books A, B, C are the only Distar books in print.

   e. Books A, B, C were selected because they corresponded to the curriculum during the 6-month period studied.

2. For the study in Exercise 1 above, the estimated variance components from the analysis of "the degree to which the teacher followed the Distar format," with all facets treated as random, were $\hat{\sigma}^2_p = 0.45$; $\hat{\sigma}^2_o = 0.07$; $\hat{\sigma}^2_b = 0.10$; $\hat{\sigma}^2_{po} = 0.03$; $\hat{\sigma}^2_{pb} = 0.03$; $\hat{\sigma}^2_{ob} = 0.04$; $\hat{\sigma}^2_{pob,e} = 0.35$. If books are considered fixed, is it preferable to conduct a separate G analysis for each book separately or to average over books?

3. Calculate the variance components for the study in Exercise 2 above, averaging over books.

4. In a study of basic skills in reading, seventh-grade students were administered a reading test with three subtests: decoding skills, vocabulary, and comprehension. Each subtest had 20 items. A portion of the data set is given in Table 5.8 (5 items per subtest for 10 students).

   a. Calculate the variance components, treating all facets as random.

   b. If subtest is a fixed facet, is it appropriate to average over subtests or to analyze each subtest separately?

   c. Treating subtest as a fixed facet, calculate the variance components for the average over subtests and for each subtest separately.

TABLE 5.8

Person	Item	Decoding 1	2	3	4	5	Vocabulary 1	2	3	4	5	Comprehension 1	2	3	4	5
1		1	0	1	1	1	1	1	1	1	0	1	0	0	0	0
2		0	0	1	1	1	1	1	1	1	0	1	0	0	1	1
3		1	1	1	1	1	1	1	1	1	0	1	1	1	1	1
4		1	0	0	1	0	1	1	1	0	0	1	0	0	0	0
5		1	1	1	1	1	1	1	1	1	1	1	1	1	1	1
6		0	1	0	1	1	1	0	1	0	1	1	1	1	1	0
7		1	0	0	0	0	1	1	0	0	0	0	0	0	0	0
8		1	1	1	1	1	1	1	1	1	1	1	0	0	0	0
9		1	0	0	0	0	0	0	0	0	0	0	0	0	0	0
10		1	1	0	1	0	1	1	1	0	0	1	1	0	0	0

Answers to Exercises

1. a. Fixed. It is not reasonable to generalize to other Distar books.
   b. If they are considered exchangeable with other Distar books, random; if not, fixed.
   c. Same as b.
   d. Fixed. Books A, B, C exhaust the universe.
   e. Same as b.

2. If books A, B, C were selected because they have specific characteristics unique to each, then a separate G study with $p \times o$ random should be conducted for each book.

   If it makes conceptual sense to average over the books, the decision should be made on the basis of the variation due to books. About 16% of the total variance is due to variability of books (b, pb, and ob effects). This fairly large percentage suggests that teacher behavior differs across books and that books should be analyzed separately. If one is interested only in the relative standing of teachers (a relative decision, see chapter 6), $\hat{\sigma}^2_b$ and $\hat{\sigma}^2_{ob}$ are not relevant, leaving the small $\hat{\sigma}^2_{pb}$. In this case, it may be reasonable to average over books.

3. $$\hat{\sigma}^2_{p*} = \hat{\sigma}^2_p + \frac{1}{n_b}\hat{\sigma}^2_{pb} = 0.45 + \frac{1}{3}\,0.03 = 0.46$$

   $$\hat{\sigma}^2_{o*} = \hat{\sigma}^2_o + \frac{1}{n_b}\hat{\sigma}^2_{ob} = 0.07 + \frac{1}{3}\,0.04 = 0.08$$

   $$\hat{\sigma}^2_{po,e*} = \hat{\sigma}^2_{po} + \frac{1}{n_b}\hat{\sigma}^2_{pob,e} = 0.03 + \frac{1}{3}\,0.35 = 0.15$$

4. a.

	p	s	i,is	ps	pi,pis,e
$\hat{\sigma}^2$ =	0.06652	0.00096	0.03593	0.01237	0.13741
% =	26	0	14	5	54

b. Because the effects of subtest are small, either strategy is defensible. If interest lies in an average reading score, averaging over subtests is appropriate. If interest lies in obtaining scores for each skill area (e.g., for diagnostic purposes), each subtest should be analyzed separately.

c. See Table 5.9.

TABLE 5.9

		Subtest		
Source of Variance	*Average over Subtests*	*Decoding*	*Vocabulary*	*Comprehension*
Persons (p)	0.07064	0.06111	0.06333	0.11222
Items (i)	0.01198	0.00556	0.05000	0.05222
pi,e	0.04580	0.17444	0.13000	0.10778

Note

1. For details underlying the statistical model, see Brennan (1983), Cronbach et al. (1972), and Erlich and Shavelson (1976).

6

Decision Studies: Measurement Error and Generalizability Coefficients

Chapters 3 through 5 have focused on generalizability studies. These studies estimate variance components for as large a universe of admissible observations as possible. These variance components are the "stuff" out of which particular measures are constructed for substantive research or decision-making studies (e.g., policy studies). Generalizability theory, then, distinguishes between *generalizability* (G) studies and *decision* (D) studies. G studies estimate the magnitude of as many potential sources of measurement error as possible. D studies use information from a G study to design a measurement that minimizes error for a *particular* purpose. The G study is associated with the development of a measurement procedure, whereas the D study applies the procedure.

A G study is designed to estimate variance components underlying a measurement procedure by defining the universe of admissible observations as broadly as possible. Once data have been collected, a large set of variance components can be estimated. These estimated variance components can be used to tailor the design of the decision study in two ways. First, the results of the G study can be used to optimize the number of conditions of each facet to obtain a desired level of reliability (generalizability), much like the Spearman–Brown prophecy formula can be used to determine appropriate test length in classical test theory. The optimal number of conditions of a facet may be less than, equal to, or greater than the number of conditions in the G study. Second, D studies can be tailored also by considering a wide variety of designs, including crossed, partially nested, and completely nested designs, fully random designs, and designs with one or more fixed facets. This chapter defines measurement error and generalizability (reliability) coefficients for different ways the data will be interpreted in the D study (relative vs. absolute interpretations).

Relative and Absolute Interpretations

G theory distinguishes between decisions based on the relative standing or ranking of individuals (*relative* interpretations) and decisions based on the absolute level of their scores (*absolute* interpretations). The way test scores are used in college admission is an example of a relative decision. Typically, a freshman class is filled with the top applicants (usually a fixed number) from the available pool of applicants. Correlating test scores with socioeconomic status is another example of a relative interpretation. Correlations are affected by the relative standing of individuals, not by their absolute levels of performance.

In contrast, the pass–fail decision based on a written examination for a driver's license is an example of an absolute decision. A score is interpreted by its level (absolute number of items correct), not by how well the person did relative to others taking the exam.

A decision maker may be concerned with one or both kinds of decisions, even for the same set of scores. The decision maker may want to know the correlation between achievement scores and, say, school attendance (a relative interpretation). Or interest may lie in assigning all students who have attained a certain level of mastery to an advanced class (an absolute decision). These different interpretations of measurements affect the definitions of error and generalizability (reliability) coefficients.

Measurement Error for Relative and Absolute Decisions

The decision maker wants to generalize from the observed score on a sample of measurements to the universe score. The inaccuracy of this generalization is called measurement error. The variance components contributing to measurement error are somewhat different for relative and absolute decisions. For *relative decisions* all variance components that influence the relative standing of individuals contribute to error. These components are the interactions of each facet with the object of measurement, here persons. For *absolute decisions* all variance components except the object of measurement contribute to measurement error. These components include all interactions and the facet main effects. Facet main effects and facet interactions index the variability in the level of scores due to differences in the difficulty or stringency of the facet conditions.

One-Facet Crossed Design

Consider the one-facet, persons-by-items ($p \times i$) science achievement test example. With persons as the object of measurement, the only variance component contributing to *relative error* is that representing the interaction between persons and items ($\sigma^2_{pi,e}$). The interaction between persons and items clearly influences persons' relative standings. A large variance component for this interaction indicates that the relative standing of persons changes from one item to another.

The variance component for items (σ^2_i), however, does *not* affect the relative standing of persons in a crossed design, in which all persons get the same items. It indicates only whether items vary in difficulty, averaging over all persons. Items may vary in difficulty (a large σ^2_i) but still preserve the relative standing of persons (a small $\sigma^2_{pi,e}$) because the item effect is independent of the $p \times i$ interaction effect. Hence, both components may be large, both may be small, or one may be large and the other small.

For absolute decisions, in contrast, all variance components except for universe-score variance contribute to measurement error. In our one-facet design the variance components contributing to *absolute error* are σ^2_i and $\sigma^2_{pi,e}$. Error for absolute decisions reflects differences in mean scores across items (σ^2_i), as well as disagreements about the relative standing of persons ($\sigma^2_{pi,e}$). When the decision maker is concerned with the absolute level of performance, differences among item means (difficulty level) shown by the magnitude of σ^2_i, affect the decision. If some items are more difficult than others, the choice of items to put on the test will influence each person's absolute level of performance. Choosing easy items would lead to high scores, averaging over persons; choosing difficult items would lead to low scores. Information about the relative standing of persons (shown by the magnitude of $\sigma^2_{pi,e}$) also influences persons' absolute scores. If the relative standing of persons changes across items, individuals' absolute scores would depend on the items chosen.

The Venn diagrams depicting the sources of variation contributing to error for relative and absolute decisions for the one-facet, crossed

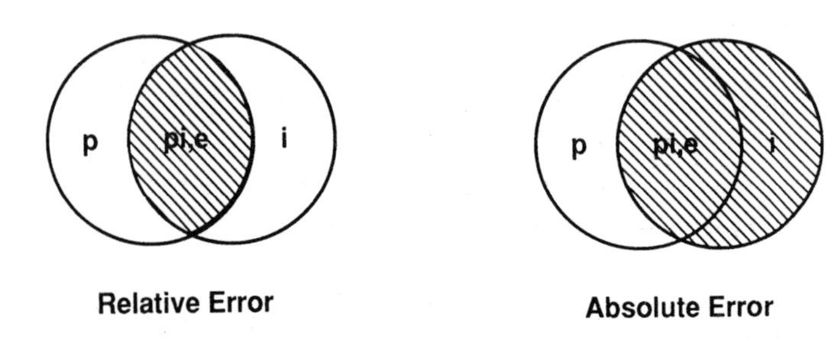

Figure 6.1. Sources of Error for Relative and Absolute Decisions for a Random $p \times i$ Design

design appear in Figure 6.1. The general rules of thumb for defining sources of relative and absolute error variance are the following: All variance components in the "p" (person) circle (the object of measurement) except the variance component for persons (universe-score variance) contribute to error for relative decisions. In a one-facet, crossed design, then, the only variance component that contributes to relative error is $\sigma^2_{pi,e}$. All variance components in the entire design except universe-score variance contribute to error for absolute decisions. The variance components contributing to absolute error for a one-facet, crossed design, then, are σ^2_i and $\sigma^2_{pi,e}$.

We can formalize the notion of *error variance* for relative and absolute decisions. For the one-facet, crossed $p \times i$ design, estimated relative error variance is

$$\sigma^2_{\text{Rel}} = \frac{\sigma^2_{pi,e}}{n'_i} \qquad [6.1]$$

and estimated absolute error variance is

$$\sigma^2_{\text{Abs}} = \frac{\sigma^2_i}{n'_i} + \frac{\sigma^2_{pi,e}}{n'_i} \qquad [6.2]$$

In the above equations, n'_i refers to the number of items to be used in the D study (not necessarily the same as the number of items, n_i, used in the G study). These expressions for error variance give the magnitude of error variance when generalizing from the average over n_i' items to the universe score. By allowing n_i' to take on different values, the decision maker can estimate the magnitude of error variance for the average over different numbers of items (different length tests). By setting n_i' equal to 10, for example, the decision maker estimates the error variance for the average score over 10 items, that is, for a 10-item test.

In Equations 6.1 and 6.2 the variance components based on a single item (σ_i^2 and $\sigma_{pi,e}^2$) are divided by n_i' to reflect the fact that the variance is a variance of means based on n_i' items.[1] This is analogous to the Spearman–Brown prophecy formula in classical theory.

Note that the residual variance component $\sigma_{pi,e}^2$ is divided only by n_i', not by the number of persons, n_p'. In G theory we are interested usually in the dependability of scores of individuals; that is, we want to know the generalizability of an individual's observed score to his or her universe score. Dividing by the number of persons n_p' would yield information about the generalizability of *group mean scores,* not individuals' scores. While the generalizability of group mean scores can be addressed in generalizability theory, this is a topic that goes beyond the development of G theory in this *Primer* (see Kane & Brennan, 1977).

Equations 6.1 and 6.2 show that the decision maker can reduce error variance by increasing the number of items and then using the average (or sum) over the items as the person's score. The greater the number of items entering into a person's score, the greater is the reduction in error variance.

Two-Facet Crossed Design

Now consider the two-facet observational study of children's help-seeking in mathematics classwork. In this study, persons were crossed with raters and occasions ($p \times r \times o$). In the original study Kenderski (1983) was interested primarily in determining whether the frequency with which children interacted with each other was related to their "cooperativeness" as measured by a questionnaire administered before group work began. This correlational question calls for a *relative interpretation* based on measurements. The magnitude of the correlation is influenced by the relative standing of children on their behavior, not by their absolute level of behavior.

Kenderski was interested also in assessing the absolute level of children's success in obtaining help. If students often failed to seek help when they needed it, she was going to recommend that children receive special training to become more assertive and to recognize their own need for help. This is an *absolute decision;* it depends on the absolute level of student behavior.

For *relative decisions* all variance components representing interactions with the object of measurement contribute to error. For this two-facet design, then, the variance components contributing to error are σ^2_{pr}, σ^2_{po}, and $\sigma^2_{pro,e}$. These variance components concern the relative standing of persons. If σ^2_{pr} is large, then raters give different relative standings of persons, and the choice of raters would affect the scores. If σ^2_{po} is large, then the relative standing of children changes across occasions, and the choice of occasions would influence the scores. If the residual variance component ($\sigma^2_{pro,e}$) is large, the relative standing of children changes across each combination of rater and occasion, and the choice of rater–occasion combinations would influence the scores.

The variance components for raters (σ^2_r), occasions (σ^2_o), and their interaction (σ^2_{ro}) do not contribute to relative error in a crossed design because they do not influence the relative standing of children. For example, children may, on average, interact more on one occasion than on another and yet have the same relative standing on both occasions.

For *absolute decisions* all variance components except the universe-score variance component contribute to error. For the two-facet design, these variance components are σ^2_r, σ^2_o, σ^2_{pr}, σ^2_{po}, σ^2_{ro}, $\sigma^2_{pro,e}$. The variance components for constant errors—raters, occasions, and their interaction—do contribute to error for absolute decisions. If some raters "see" more help-seeking than other raters, averaging over children and occasions (a large σ^2_r), the particular raters sampled will affect the decision about the prevalence of this sequence of behavior. Similarly, if children seek more help on one occasion than another, averaging over children and raters (a large σ^2_o), then the choice of occasion will influence the decision. Finally, if the measurement indicated that children seek more help with one combination of raters and occasions than another (a large σ^2_{ro}), the particular combination of raters and occasions sampled will affect the decision.

The Venn diagrams depicting the sources of variation contributing to error for relative and absolute decisions for the two-facet, crossed design appear in Figure 6.2. All variance components in the *"p"* circle

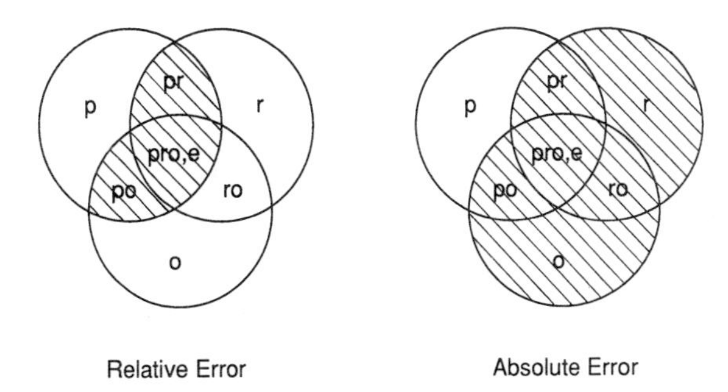

Figure 6.2. Sources of Error for Relative and Absolute Decisions for a Random $p \times r \times o$ Design

except universe-score variance contribute to relative error. All variance components in the entire design except universe-score variance contribute to absolute error.

The general formulas for relative and absolute error variance for the fully random, crossed, two-facet design are

$$\sigma^2_{\text{Rel}} = \frac{\sigma^2_{pr}}{n'_r} + \frac{\sigma^2_{po}}{n'_o} + \frac{\sigma^2_{pro,e}}{n'_r n'_o} \qquad [6.3]$$

and

$$\sigma^2_{\text{Abs}} = \frac{\sigma^2_{r}}{n'_r} + \frac{\sigma^2_{o}}{n'_o} + \frac{\sigma^2_{pr}}{n'_r} + \frac{\sigma^2_{po}}{n'_o} + \frac{\sigma^2_{ro}}{n'_r n'_o} + \frac{\sigma^2_{pro,e}}{n'_r n'_o} \qquad [6.4]$$

The divisor for each variance component is the number of conditions of each error facet represented in the variance component. In the two-facet crossed design, as was true in the one-facet design, error variance can be reduced by increasing the number of conditions of a facet entering into a child's score. Increasing the number of raters (n'_r) and occasions (n'_o) on which to base a score will reduce error variance correspondingly.

Two-Facet, Partially Nested Design

In the Erlich and Shavelson (1976) study of teacher behavior, different teachers (p) were observed on different occasions (o), but all raters (r) observed all teachers on all occasions. So the design was $(o{:}p) \times r$. Both relative and absolute decisions could be of interest here. For example, interest may lie in determining the relationship between teacher behavior and student achievement (a relative interpretation). But the absolute level of teacher behavior also may be of concern. Teachers who show too little desirable behavior or too much undesirable behavior might be selected to participate in an in-service training program to modify their behavior (an absolute decision).

For relative decisions, only the variance components representing interactions with persons contribute to error. In this $(o{:}p) \times r$ design the relevant variance components are σ^2_{pr}, $\sigma^2_{o,po}$, and $\sigma^2_{ro,pro,e}$. For absolute decisions all variance components except for the universe-score variance component contribute to error. In this design the relevant variance components are σ^2_{r}, σ^2_{pr}, $\sigma^2_{o,po}$, and $\sigma^2_{ro,pro,e}$.

The Venn diagrams depicting sources of variation contributing to error for relative and absolute decisions for the $(o{:}p) \times r$ design appear in Figure 6.3. For the two-facet, partially nested $(o{:}p) \times r$ design, the formulas for relative and absolute error variance are

$$\sigma^2_{\text{Rel}} = \frac{\sigma^2_{pr}}{n'_r} + \frac{\sigma^2_{o,po}}{n'_o} + \frac{\sigma^2_{ro,pro,e}}{n'_r n'_o} \qquad [6.5]$$

$$\sigma^2_{\text{Abs}} = \frac{\sigma^2_{r}}{n'_r} + \frac{\sigma^2_{pr}}{n'_r} + \frac{\sigma^2_{o,po}}{n'_o} + \frac{\sigma^2_{ro,pro,e}}{n'_r n'_o} \qquad [6.6]$$

Designs with a Fixed Facet

The previous chapter showed how to compute variance components for designs with a fixed facet. Consider the study of teacher behavior in reading and mathematics. Teachers (p) were observed by multiple raters (r) in both subject matters (s). The design was fully crossed: $p \times r \times s$. Subject matter was a fixed facet. Chapter 5 discussed two ways to analyze designs with a fixed facet: averaging over subject matters, and analyzing each subject matter separately. Both procedures leave a

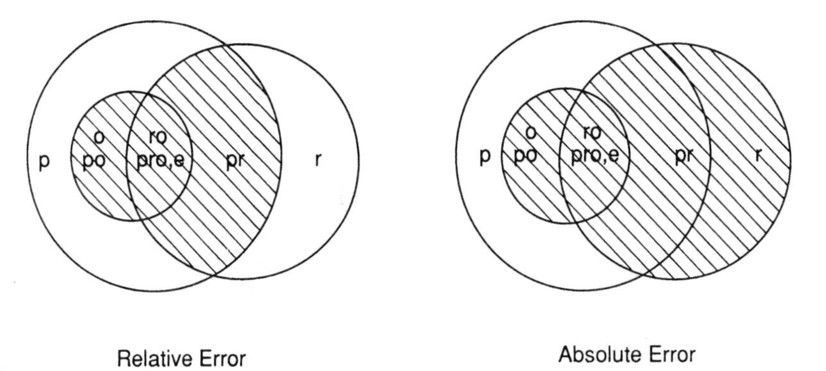

Figure 6.3. Sources of Error for Relative and Absolute Decisions for a Random $(o{:}p) \times r$ Design

fully crossed design with persons crossed with raters. Averaging over subject matters gives estimated variance components denoted as σ^2_{p*}, σ^2_{r*}, and $\sigma^2_{pr,e*}$. The formulas for relative and absolute error variance are

$$\sigma^2_{\text{Rel}} = \frac{\sigma^2_{pr,e*}}{n'_r} \qquad [6.7]$$

$$\sigma^2_{\text{Abs}} = \frac{\sigma^2_{r*}}{n'_r} + \frac{\sigma^2_{pr,e*}}{n'_r} \qquad [6.8]$$

Analyzing each subject matter separately gives variance components for each subject matter denoted as σ^2_p, σ^2_r, and $\sigma^2_{pr,e}$. The formulas for relative and absolute error variance are

$$\sigma^2_{\text{Rel}} = \frac{\sigma^2_{pr,e}}{n'_r} \qquad [6.9]$$

$$\sigma^2_{\text{Abs}} = \frac{\sigma^2_r}{n'_r} + \frac{\sigma^2_{pr,e}}{n'_r} \qquad [6.10]$$

For the generalizability study of the Self-Description Questionnaire (SDQ), persons (p) were administered three self-concept scales (s), each with multiple items (i). The items were different from scale to scale, so items were nested within scales. The design was $p \times (i{:}s)$, with scale considered a fixed facet. Averaging over conditions of the fixed facet s and analyzing each scale separately, both leave a fully random $p \times i$ design. The variance components when averaging over scales are denoted as σ^2_{p*}, σ^2_{i*}, and $\sigma^2_{pi,e*}$. The formulas for relative and absolute error variance are

$$\sigma^2_{\text{Rel}} = \frac{\sigma^2_{pi,e*}}{n'_i} \qquad [6.11]$$

$$\sigma^2_{\text{Abs}} = \frac{\sigma^2_{i*}}{n'_i} + \frac{\sigma^2_{pi,e*}}{n'_i} \qquad [6.12]$$

When analyzing each scale separately, the variance components are σ^2_p, σ^2_i, and $\sigma^2_{pi,e}$. The formulas for relative and absolute error variance are

$$\sigma^2_{\text{Rel}} = \frac{\sigma^2_{pi,e}}{n'_i} \qquad [6.13]$$

$$\sigma^2_{\text{Abs}} = \frac{\sigma^2_i}{n'_i} + \frac{\sigma^2_{pi,e}}{n'_i} \qquad [6.14]$$

The same principles for defining relative and absolute error variance can be applied to any other design, as shown in Appendix 6.1.

Generalizability Coefficients for Relative and Absolute Decisions

Although G theory stresses the importance of variance components and measurement error, it also provides a generalizability coefficient that is analogous to the reliability coefficient in classical theory, namely, true-score variance divided by expected observed-score variance (i.e., an intraclass correlation coefficient). For relative decisions the G coefficient for a single-facet design is

$$E\rho^2_{\text{Rel}} = \frac{\sigma^2_p}{E\sigma^2(X_{pi})} = \frac{\sigma^2_p}{(\sigma^2_p + \sigma^2_{\text{Rel}})} \qquad [6.15]$$

The notation $E\rho^2$ (Cronbach et al., 1972) is meant to show that "a generalizability coefficient is approximately equal to the expected value . . . of the squared correlation between observed scores and universe scores" (Brennan, 1983, p. 17). Note that the expected observed-score variance, $E\sigma^2 X_{pi}$, is not the same as the total variance, $\sigma^2 X_{pi}$. The expected observed-score variance is defined as

$$E\sigma^2(X_{pi}) = E_i E_p (X_{pi} - \mu_i)^2 = \sigma^2_p + \sigma^2_{pi,e} \qquad [6.16]$$

The expected observed-score variance reflects variability in the ordering of observations (relative standing). It constitutes the entire *p* circle in a Venn diagram (Figure 6.1) and is the sum of the universe-score variance plus relative-error variance. Contrast this with the total variance: $\sigma^2_p + \sigma^2_i + \sigma^2_{pi,e}$. The total variance is a measure of the total variation in item responses and reflects variability in absolute level of performance.

A reliability-like coefficient also can be defined for absolute decisions. Brennan and Kane (1977) called this coefficient an *index of dependability* and used the symbol ϕ (phi):

$$\phi = \frac{\sigma^2_p}{(\sigma^2_p + \sigma^2_{\text{Abs}})} \qquad [6.17]$$

Because the denominator is not the expected observed-score variance and the coefficient does not approximate the expected value of the squared correlation between observed and universe scores, it is not, strictly speaking, appropriate to call phi a generalizability coefficient or to use the symbol $E\rho^2_{\text{Abs}}$. So we follow the convention of Brennan and Kane (1977) in using the symbol phi. If the generalizability coefficient is viewed more broadly as the universe-score variance divided by the sum of universe score variance plus error variance, then the phi coefficient can be seen as a generalizability coefficient for absolute decisions.

Comparison with Classical Test Theory

Classical test theory assumes strictly parallel measurements; that is, the means across items are assumed to be equal, as are the variances (but see Lord & Novick, 1968). Hence, in a $p \times i$ measurement, the item effect is assumed to be zero. Because classical theory assumes no item variability, it cannot be used when the investigator's interest is in making absolute decisions concerning tests that are not strictly parallel. This limitation applies to all designs.

A consequence of the parallel-measurements assumption is that classical test theory is primarily a theory of individual differences; that is, it usually is concerned with the relative standing of individuals. The reliability coefficient, following this individual differences focus, is calculated only with sources of variation for persons and the residual:

$$\rho_{xx'} = \frac{\sigma_p^2}{\sigma_p^2 + \dfrac{\sigma_{pi,e}^2}{n_i'}} \qquad [6.18]$$

For a one-facet design, then, the reliability coefficient in classical theory is comparable to the generalizability coefficient for relative decisions. This is true only for designs with one random facet as will be seen below.

Because classical theory partitions variance into only two sources, true-score and error variance, it can estimate only one source of error at a time. For the two-facet study with persons crossed with raters and occasions, classical theory would have to examine separately consistency over raters and stability over time. For calculating consistency over raters, the analysis in classical theory could be done separately for each occasion; that is, the two-facet persons-by-raters-by-occasions design could be separated into multiple one-facet designs: a persons-by-raters design for each occasion. A reliability coefficient for raters, analogous to that in Equation 6.18, then could be calculated at each occasion. These reliability coefficients, intraclass correlation coefficients, would indicate the consistency among the raters' scores on the various occasions. Alternatively, the scores across multiple occasions could be averaged, producing a single one-facet design and a reliability coefficient for raters averaged over time. Similar analyses could be performed to examine the stability of persons' behavior over occasions for each rater one at a time, or averaged over raters.

Two drawbacks are evident with this approach. The primary drawback is that classical theory cannot disentangle effects of error variation due to raters and occasions in a single analysis. The second drawback is that the approaches available in classical theory typically address only reliability for relative decisions. In this two-facet design, classical theory assumes that the rater and occasion effects are zero; that is, that the variance components for raters, occasions, and the interactions between raters and occasions are all zero. G theory solves both of these problems.

Summary

G theory distinguishes between relative and absolute interpretations of behavioral measurements. Relative interpretations address decisions about "how much better" one individual performed than another. Absolute interpretations address decisions about "how well" an individual can perform, regardless of the performance of his or her peers. The way in which a measurement is interpreted directly influences the definition of measurement error and the definition of the generalizability (reliability) coefficient.

For relative decisions measurement error is defined as all variance components that influence the relative standing of individuals. These components are the interactions of each facet with the object of measurement, usually persons. For absolute decisions all variance components except the object of measurement contribute to measurement error. Not only do changes in the ranking of individuals contribute to error, the actual levels of their performances also depend on whether the items are easy or difficult or whether the raters are lenient or strict. In a one-facet, person-by-items measurement, relative error is defined as the interaction variance of persons with items confounded with other sources: $\sigma^2_{\text{Rel}} = \sigma^2_{pi,e}$. Absolute error is defined as both the interaction variance and the variance component for items (variability arising from differences in difficulty from one item to another): $\sigma^2_{\text{Abs}} = \sigma^2_{pi,e} + \sigma^2_i$. This reasoning readily extends to multifacet designs with nested and/or fixed facets.

Although it focuses on variance components, G theory also has a reliability coefficient, analogous to the reliability coefficient in classical test theory, called a generalizability coefficient. The G coefficient is defined as the universe-score variance divided by the expected observed-score variance: $E\rho^2_{\text{Rel}} = \sigma^2_p / (\sigma^2_p + \sigma^2_{\text{Rel}})$. The G coefficient, then, is the proportion of expected observed-score variance that is universe-score variance. G theory also defines a coefficient that reflects the reliability of a measurement used for absolute decisions, called the index of dependability. It is the proportion of total variance that is accounted for by universe-score variance: $\phi = \sigma^2_p / (\sigma^2_p + \sigma^2_{\text{Abs}})$. This reasoning for defining coefficients of generalizability and dependability extends to multifacet measurements with nested and/or fixed facets.

Appendix 6.1

Relative and Absolute Error Variance for Two-Facet Random Designs

$p \times i \times j$ Design

$$\sigma^2_{\text{Rel}} = \frac{\sigma^2_{pi}}{n'_i} + \frac{\sigma^2_{pj}}{n'_j} + \frac{\sigma^2_{pij,e}}{n'_i n'_j}$$

$$\sigma^2_{\text{Abs}} = \frac{\sigma^2_{i}}{n'_i} + \frac{\sigma^2_{j}}{n'_j} + \frac{\sigma^2_{ij}}{n'_i n'_j} + \frac{\sigma^2_{pi}}{n'_i} + \frac{\sigma^2_{pj}}{n'_j} + \frac{\sigma^2_{pij,e}}{n'_i n'_j}$$

$p \times (i{:}j)$ Design

$$\sigma^2_{\text{Rel}} = \frac{\sigma^2_{pj}}{n'_j} + \frac{\sigma^2_{pi,pij,e}}{n'_i n'_j}$$

$$\sigma^2_{\text{Abs}} = \frac{\sigma^2_{j}}{n'_j} + \frac{\sigma^2_{i,ij}}{n'_i n'_j} + \frac{\sigma^2_{pj}}{n'_j} + \frac{\sigma^2_{pi,pij,e}}{n'_i n'_j}$$

$(i{:}p) \times j$ Design

$$\sigma^2_{\text{Rel}} = \frac{\sigma^2_{pj}}{n'_j} + \frac{\sigma^2_{i,pi}}{n'_i} + \frac{\sigma^2_{ij,pij,e}}{n'_i n'_j}$$

$$\sigma^2_{\text{Abs}} = \frac{\sigma^2_{j}}{n'_j} + \frac{\sigma^2_{pj}}{n'_j} + \frac{\sigma^2_{i,pi}}{n'_i} + \frac{\sigma^2_{ij,pij,e}}{n'_i n'_j}$$

i:($p \times j$) Design

$$\sigma^2_{\text{Rel}} = \frac{\sigma^2_{pj}}{n'_j} + \frac{\sigma^2_{i,pi,ij,pij,e}}{n'_i n'_j}$$

$$\sigma^2_{\text{Abs}} = \frac{\sigma^2_j}{n'_j} + \frac{\sigma^2_{pj}}{n'_j} + \frac{\sigma^2_{i,pi,ij,pij,e}}{n'_i n'_j}$$

($i \times j$):p Design

$$\sigma^2_{\text{Rel}} = \frac{\sigma^2_{i,pi}}{n'_i} + \frac{\sigma^2_{j,pj}}{n'_j} + \frac{\sigma^2_{ij,pij,e}}{n'_i n'_j}$$

$$\sigma^2_{\text{Abs}} = \frac{\sigma^2_{i,pi}}{n'_i} + \frac{\sigma^2_{j,pj}}{n'_j} + \frac{\sigma^2_{ij,pij,e}}{n'_i n'_j}$$

i:j:p Design

$$\sigma^2_{\text{Rel}} = \frac{\sigma^2_{j,pj}}{n'_j} + \frac{\sigma^2_{i,pi,ij,pij,e}}{n'_i n'_j}$$

$$\sigma^2_{\text{Abs}} = \frac{\sigma^2_{j,pj}}{n'_j} + \frac{\sigma^2_{i,pi,ij,pij,e}}{n'_i n'_j}$$

Exercises

1. The estimated variance components for the G study of person perception and attribution (see chapter 3, Exercise 1) are $\hat{\sigma}^2_p = 0.025$, $\hat{\sigma}^2_i = 0.050$, $\hat{\sigma}^2_{pi,e} = 0.171$. Calculate the estimated error variances and G coefficients for relative and absolute decisions for a one-item test.
2. The estimated variance components for the hypothetical G study of vocational interest are given in Table 6.1 (see chapter 3, Exercise 3). Calculate the estimated error variances and G coefficients for relative and absolute decisions for a D study with one item and one occasion.
3. The estimated variance components for the study of Marine Corps infantry riflemen evaluated while assembling communications equipment are given in Table 6.2 (see chapter 4, Exercise 3). Calculate the estimated variance components and G coefficients for relative and absolute decisions for a D study with one occasion and one rater.

TABLE 6.1

Source of Variation	$\hat{\sigma}^2$
Persons (p)	0.5333
Items (i)	0.1683
Occasions (o)	0.0133
pi	0.4733
po	0.0217
io	0[a]
pio,e	0.5583

a. Negative estimated variance component set equal to zero.

TABLE 6.2

Source of Variation	$\hat{\sigma}^2$
Persons (p)	0.0125
Occasions (o)	0.0070
Raters:Occasions ($r{:}o$)	0.0013
po	0.0035
pr,pro,e	0.0154

Answers to Exercises

1. $\hat{\sigma}^2_{\text{Rel}} = 0.171$; $\hat{\sigma}^2_{\text{Abs}} = 0.221$; $\hat{\rho}^2 = 0.13$; $\hat{\phi} = 0.10$
2. $\hat{\sigma}^2_{\text{Rel}} = 1.0533$; $\hat{\sigma}^2_{\text{Abs}} = 1.2349$; $\hat{\rho}^2 = 0.34$; $\hat{\phi} = 0.30$
3. $\hat{\sigma}^2_{\text{Rel}} = 0.0189$; $\hat{\sigma}^2_{\text{Abs}} = 0.0272$; $\hat{\rho}^2 = 0.40$; $\hat{\phi} = 0.31$

Note

1. This formalism is analogous to the relation between the variance σ^2 and the error variance σ^2 / n (defined as the standard error of the mean). The former refers to a single score, and the latter refers to the variability among sample means based on n observations.

7

Generalizability and Decision Studies with the Same Design

The previous chapter showed how to estimate error variance and generalizability in the decision study. This chapter shows how to use the results of the G study to optimize the number of conditions of each facet in the design of the D study. Here we focus on G studies and D studies with the same design; the next chapter shows how to use the results of a G study to estimate error variance and generalizability for D studies with different designs.

Generalizability and Decision Studies for Crossed Designs

One-Facet, Crossed $p \times i$ Design

The G study is used to estimate variance components; the D study uses these estimates to optimize the measurement for a particular decision-making purpose. To see how this works, we will begin by using, as an example, the study of science achievement for which a sample of persons (p) responded to all items (i) on the test, producing a $p \times i$ design.

Table 7.1 presents the relevant information: (a) sources of variation for the $p \times i$ design, (b) the estimated variance components from the G study, (c) the estimated variance components that are projected for alternative D studies, (d) estimated error variances for relative and absolute decisions, and (e) estimated generalizability coefficients for relative and absolute decisions.

TABLE 7.1 Decision Study for the Measurement of Science Achievement ($p \times i$ Design)

			G Study	*Alternative D Studies*			
Source of Variation	$\hat{\sigma}^2$	$n'_i =$	*1*	*8*	*20*	*30*	*40*
Persons (p)	$\hat{\sigma}^2_p$		0.0305	0.0305	0.0305	0.0305	0.0305
Items (i)	$\hat{\sigma}^2_i$		0.0093	0.0012	0.0005	0.0003	0.0002
pi,e	$\hat{\sigma}^2_{pi,e}$		0.2103	0.0263	0.0105	0.0070	0.0053
$\hat{\sigma}^2_{\text{Rel}}$			0.2103	0.0263	0.0105	0.0070	0.0053
$\hat{\sigma}^2_{\text{Abs}}$			0.2196	0.0275	0.0110	0.0073	0.0055
$\hat{\rho}^2$			0.13	0.54	0.74	0.81	0.85
$\hat{\phi}$			0.12	0.53	0.73	0.81	0.85

The first column of numbers in Table 7.1 are the estimated variance components produced by the G study. These variance components show the variability of all sources of variation based on sampling a single item from the item universe. Relative and absolute error variance for a one-item test are 0.2103 and 0.2196, respectively. These show the magnitude of error when generalizing from the score on a single item to the universe score. If only one item were used to measure science achievement, the generalizability and phi coefficients would be 0.13 and 0.12, respectively, clearly too low for any decision-making purpose.

Table 7.1 also gives estimated variance components, estimated error variances, and estimated generalizability and phi coefficients for different numbers of items. These estimates are calculated in the following way. First, the variance components from the G study serve as estimates of parameters of interest (σ^2_p, σ^2_i, $\sigma^2_{pi,e}$). Second, each variance component contributing to relative or absolute measurement error (or both) is divided by the number of items n'_i that might be used in a D study. This follows from the definition of measurement error in a one-facet, crossed design:[1]

$$\hat{\sigma}^2_{\text{Rel}} = \frac{\hat{\sigma}^2_{pi,e}}{n'_i} \qquad [7.1]$$

$$\hat{\sigma}^2_{\text{Abs}} = \frac{\hat{\sigma}^2_i}{n'_i} + \frac{\hat{\sigma}^2_{pi,e}}{n'_i} \quad [7.2]$$

For example, to get the estimated variance components for a D study with an eight-item test, we divide $\hat{\sigma}^2_i$ and $\hat{\sigma}^2_{pi,e}$ by $n'_i = 8$, to get 0.0012 and 0.0263, respectively.

Even using the average over eight items for an achievement test score (the number of items used in the generalizability study) yields mediocre levels of estimated generalizability: 0.54 and 0.53 for relative and absolute decisions, respectively. About 30 items are needed to obtain reasonable levels of estimated generalizability, say, about 0.80.

As can be seen in Table 7.1, each increase of 10 items yields a smaller increase in estimated levels of generalizability. At some point the gain in estimated generalizability will not be worth the expense of developing additional items and the cost and time needed to administer them.

In this example, estimated error variances and generalizability coefficients are very similar for relative and absolute decisions. The variance component for items enters into error variance for absolute decisions but not for relative decisions. Because the variability due to items is so small here, its inclusion or exclusion has little impact on error variance. If the items had varied more in difficulty, the difference between error variances and, hence, the difference between the generalizability and phi coefficients would have been greater.

Two-Facet, Crossed $p \times r \times o$ Design

To show how to use G-study information to project alternative D-study measurements with two crossed facets, we use the study of children's help-seeking behavior in which 13 children were tape-recorded while working on mathematics problems on two occasions. Two raters coded all tapes, counting the number of times each child asked for help. The design has persons (p) crossed with raters (r) and occasions (o). All sources of variance are considered random. Table 7.2 presents the variance components, error variances, and generalizability coefficients for the $p \times r \times o$ design.

The first column of estimated variance components in Table 7.2 are those produced by the G study. If only one occasion and one rater were used, the generalizability and phi coefficients would be 0.40 and 0.35, respectively. These levels of reliability are too low for decision-

TABLE 7.2 Decision Study for Observations of Children's Help-Seeking Behavior ($p \times r \times o$ Design)

Source of Variation	$\hat{\sigma}^2$		G Study	Alternative D Studies			
		$n'_r =$	*1*	*1*	*2*	*4*	*4*
		$n'_o =$	*1*	*4*	*4*	*2*	*3*
Persons (p)	$\hat{\sigma}^2_p$		0.3974	0.3974	0.3974	0.3974	0.3974
Raters (r)	$\hat{\sigma}^2_r$		0.0096	0.0096	0.0048	0.0024	0.0024
Occasions (o)	$\hat{\sigma}^2_o$		0.1090	0.0273	0.0273	0.0545	0.0363
pr	$\hat{\sigma}^2_{pr}$		0.0673	0.0673	0.0337	0.0168	0.0168
po	$\hat{\sigma}^2_{po}$		0.3141	0.0785	0.0785	0.1571	0.1047
ro	$\hat{\sigma}^2_{ro}$		0.0064	0.0016	0.0008	0.0008	0.0005
pro,e	$\hat{\sigma}^2_{pro,e}$		0.2244	0.0561	0.0281	0.0281	0.0187
$\hat{\sigma}^2_{\text{Rel}}$			0.6058	0.2019	0.1403	0.2020	0.1402
$\hat{\sigma}^2_{\text{Abs}}$			0.7308	0.2404	0.1732	0.2597	0.1794
$\hat{\rho}^2$			0.40	0.66	0.74	0.66	0.74
$\hat{\phi}$			0.35	0.62	0.70	0.60	0.69

making purposes. Increasing the number of raters and/or occasions would reduce error variance and, consequently, increase the level of generalizability.

To see the effects of increasing the number of raters and occasions, we divide each potential estimated error-variance component by n'_r, n'_o, or $n'_r n'_o$ based on the definition of relative and absolute error given in chapter 6:

$$\hat{\sigma}^2_{\text{Rel}} = \frac{\hat{\sigma}^2_{pr}}{n'_r} + \frac{\hat{\sigma}^2_{po}}{n'_o} + \frac{\hat{\sigma}^2_{pro,e}}{n'_r n'_o} \qquad [7.3]$$

$$\hat{\sigma}^2_{\text{Abs}} = \frac{\hat{\sigma}^2_{r}}{n'_r} + \frac{\hat{\sigma}^2_{o}}{n'_o} + \frac{\hat{\sigma}^2_{pr}}{n'_r} + \frac{\hat{\sigma}^2_{po}}{n'_o} + \frac{\hat{\sigma}^2_{ro}}{n'_r n'_o} + \frac{\hat{\sigma}^2_{pro,e}}{n'_r n'_o} \qquad [7.4]$$

The choice of number of raters and number of occasions depends on the magnitude of measurement error, on the estimated level of generalizability desired, and on practical considerations. Because the variability due to occasions is greater than that for raters, increasing the number of occasions will have a larger effect on the estimated error variance and level of generalizability than will increasing the number of raters (see Table 7.2). For example, using four occasions and two raters gives a smaller error variance and greater generalizability and phi coefficients than does using two occasions and four raters.

Practical considerations also may guide the choice of number of occasions and raters. It is usually more costly (time, money, logistics) to observe persons on multiple occasions than to train multiple raters. If the decison maker wishes to observe (in this case, tape-record) persons on only three occasions, it would be necessary to use four raters to reach a generalizability coefficient (relative decision) of 0.74. This is a total of 12 ratings per person, in contrast to 8 ratings if persons were observed four times using two raters. But the 12 ratings may prove less costly than the 8 ratings even though they yield similar estimated generalizability coefficients.

The decision maker should take into account the fact that increasing the number of raters and occasions beyond a certain point yields diminishing returns. For four occasions, using one rater gives an estimated phi coefficient of 0.62 (see Table 7.2). Increasing the number of raters to two yields an estimated phi coefficient of 0.70. To reach a phi coefficient of 0.78, it would take 22 raters, clearly impractical!

Generalizability and Decision Studies for Other Designs

The use of G study information in projecting alternative D-study designs extends to more complex measurements. Here we show how G-study information can be used to design D studies for measurements with nested and fixed facets.

One-Facet Nested Design

The use of G-study information to project alternative D-study designs for nested measurements parallels that for crossed measurements. Consider the hypothetical study of mathematics achievement described in

TABLE 7.3 Decision Study for the Mathematics Achievement Test (i:p Design)

Source of Variation	$\hat{\sigma}^2$	n_i' =	*G Study* 1	*Alternative D Studies* 10	15	20
Persons (p)	$\hat{\sigma}^2_p$		0.0750	0.0750	0.0750	0.0750
Items: Persons (i:p)	$\hat{\sigma}^2_{i,pi,e}$		0.2000	0.0200	0.0133	0.0100
$\hat{\sigma}^2_{\text{Rel}}$			0.2000	0.0200	0.0133	0.0100
$\hat{\sigma}^2_{\text{Abs}}$			0.2000	0.0200	0.0133	0.0100
$\hat{\rho}^2$			0.27	0.79	0.85	0.88
$\hat{\phi}$			0.27	0.79	0.85	0.88

chapter 4. Five persons (p) each responded to two different items (i). This produces a one-facet, nested i:p (items nested within persons) measurement. The estimated variance components, error variances, and generalizability coefficients for this measurement appear in Table 7.3. The variance components for alternative D studies were estimated in the same way as in the crossed designs described above. The variance components estimated from the G study were divided by the number of items that could go into the achievement test.

First, note that this nested measurement has only two sources of variation: universe-score variance ($\hat{\sigma}^2_p$) and one term for error ($\hat{\sigma}^2_{i,pi,e}$). In this nested design the item effect is confounded with the usual residual term and cannot be estimated separately. Consequently, the error variance for relative decisions is the same as the error variance for absolute decisions.

Because relative and absolute error variance are the same in the one-facet nested design, so too are the generalizability and phi coefficients. The coefficients show the diminishing returns for increasing the number of items per person: A 20-item test produces a fairly small increase in estimated generalizability over a 15-item test. The decision maker must balance cost considerations and test length in choosing the optimal number of items.

Note that the confounding of the item effect and the residual effect shows the disadvantage of using a nested design for a G study. We would like to know the effect of item difficulty by itself; this provides more information than does the confounded error variance. So whenever possible, crossed G-study designs should be used.

Two Facet, Partially Nested $(o{:}p) \times r$ Design

In the study of teacher behavior ("direction to try again") described earlier (Erlich & Shavelson, 1976), five teachers were videotaped on three occasions. Three raters coded all 15 videotapes. The occasions were different for each teacher, so occasions were nested within teachers. Each rater coded all teachers on all occasions, so raters were crossed with teachers and occasions. This design is partially nested, with occasions (o) nested within teachers (p) and both crossed with raters (r). The notation for this design is $(o{:}p) \times r$.

Table 7.4 gives the estimated error variances based on the G study and alternative D-study designs. Once again, the estimated error-variance components for alternative D-study designs were calculated by dividing the components estimated in the G study by the number of raters, occasions, or both, that are being considered for the D study. Equations 6.5 and 6.6 from chapter 6 can be adapted for this purpose:

$$\hat{\sigma}^2_{\text{Rel}} = \frac{\hat{\sigma}^2_{pr}}{n'_r} + \frac{\hat{\sigma}^2_{o,po}}{n'_o} + \frac{\hat{\sigma}^2_{ro,pro,e}}{n'_r n'_o} \qquad [7.5]$$

$$\hat{\sigma}^2_{\text{Abs}} = \frac{\hat{\sigma}^2_{r}}{n'_r} + \frac{\hat{\sigma}^2_{pr}}{n'_r} + \frac{\hat{\sigma}^2_{o,po}}{n'_o} + \frac{\hat{\sigma}^2_{ro,pro,e}}{n'_r n'_o} \qquad [7.6]$$

As can be seen in Table 7.4, using one rater and one occasion to measure teacher behavior would yield unacceptably low generalizability and phi coefficients: 0.35 and 0.32, respectively. Because the variability due to raters and occasions are both substantial, it is necessary to have multiple raters and occasions. A design with three raters and three occasions (a total of nine ratings per teacher in a crossed design) yields minimally acceptable generalizability coefficients. Because of cost considerations, however, some other combination might be selected. In this study, raters coded videotapes of teacher behavior, a very time-consuming (and, hence, expensive) process. Costs can be reduced by minimizing the total number of ratings made per teacher. So, the decision maker may select a design yielding eight ratings per

TABLE 7.4 Decision Study for the Study of Teacher Behavior[a] [$(o{:}p) \times r$ Design]

Source of Variation	$\hat{\sigma}^2$		*G Study*	*Alternative D Studies*			
		$n'_r =$	*1*	2	*3*	2	*4*
		$n'_o =$	*1*	3	*3*	4	*2*
Teachers (p)	$\hat{\sigma}^2_p$		8.288	8.288	8.288	8.288	8.288
Raters (r)	$\hat{\sigma}^2_r$		2.613	1.307	0.871	1.307	0.653
Occasions: Teachers ($o{:}p$)	$\hat{\sigma}^2_{o,po}$		4.622	1.541	1.541	1.156	2.311
pr	$\hat{\sigma}^2_{pr}$		1.537	0.769	0.512	0.769	0.384
o:pr,e	$\hat{\sigma}^2_{ro,pro,e}$		9.072	1.512	1.008	1.134	1.134
$\hat{\sigma}^2_{\text{Rel}}$			15.231	3.822	3.061	3.059	3.829
$\hat{\sigma}^2_{\text{Abs}}$			17.844	5.129	3.932	4.366	4.482
$\hat{\rho}^2$			0.35	0.68	0.73	0.73	0.68
$\hat{\phi}$			0.32	0.62	0.68	0.65	0.65

a. Behavior variable = "Direction to try again."

teacher, either two raters and four occasions or four raters and two occasions, instead of nine. To choose between alternatives, the decision maker must balance the cost of training raters to code videotapes, the cost of videotaping teachers on multiple occasions, and estimated generalizability.

Two-Facet Crossed $p \times r \times s$ Mixed Design with s Fixed

Consider the crossed, two-facet design with a fixed facet using the hypothetical study of teacher behavior presented in chapter 5. Eight teachers were each observed by three raters in two subject matters: reading and mathematics (cf. Erlich & Shavelson, 1978). Teachers (p) and raters (r) are random; subject matter (s) is fixed.

As was discussed in chapter 5, the decision maker has two options in analyzing designs with a fixed facet: averaging over conditions of the fixed facet, or analyzing each condition separately. The choice between these should be made on conceptual grounds, on the basis of what makes sense and how information will be used in the D study. Regardless of

TABLE 7.5 Decision Study for Teacher Behavior Observations ($p \times r \times s$ Design with s Fixed)

			Averaging over Subject Matters		*Mathematics*		*Reading*		
Source of Variation	$\hat{\sigma}^2$	$n_r' =$	*1*	2	*1*	2	*1*	2	*3*
Teachers (p)	$\hat{\sigma}_p^2$		2.3438	2.3438	3.8869	3.8869	3.1429	3.1429	3.1429
Raters (r)	$\hat{\sigma}_r^2$		0.0342	0.0171	0[a]	0[a]	0.1548	0.0774	0.0516
pr,e	$\hat{\sigma}_{pr,e}^2$		0.9450	0.4725	0.5060	0.2530	2.2619	1.1310	0.7540
$\hat{\sigma}_{\text{Rel}}^2$			0.9450	0.4725	0.5060	0.2530	2.2619	1.1310	0.7540
$\hat{\sigma}_{\text{Abs}}^2$			0.9792	0.4896	0.5060	0.2530	2.4167	1.2084	0.8056
$\hat{\rho}^2$			0.71	0.83	0.88	0.94	0.58	0.74	0.81
$\hat{\phi}$			0.71	0.83	0.88	0.94	0.57	0.72	0.80

a. Variance component of –0.0476 set equal to zero.

whether the decision maker decides to explore the generalizability of the average over the two subject matters or the generalizability of reading and mathematics separately, both cases resolve into a one-facet, teachers crossed with raters design. This resolution occurs because either the two subject matters are "averaged over" to create a $p \times r$ design, or the $p \times r$ design is applied to each subject matter separately. The variance components involving raters are divided by the number of raters in the D study, and estimated error variances and generalizability coefficients are calculated accordingly.

Table 7.5 contains estimated variance components, error variances, and generalizability coefficients for the average of the two subject matters and for each subject matter separately. Once again, the estimated error-variance components for the alternative D studies were obtained by dividing the estimated error-variance components from the G study by n_r', the number of raters being considered for alternative D studies.

When averaging over reading and mathematics, to reach an estimated level of generalizability of 0.80 two raters are needed. When subject matters are analyzed separately, the required number of raters in the D study is different in reading and mathematics. For an estimated level of

TABLE 7.6 D Study for the Self-Description Questionnaire [$p \times (i{:}s)$ Design with s Fixed]

							Scale				
		Averaging Over Scales		*General*		*Academic*			*Mathematics*		
Source of Variation	$n'_i =$	*1*	*10*	*1*	*10*	*1*	*9*	*10*	*1*	*8*	*10*
Persons (p)		0.5898	0.5898	0.6100	0.6100	0.8054	0.8054	0.8054	0.9761	0.9761	0.9761
Items (i)		0.2668	0.0267	0.1228	0.0123	0.3585	0.0398	0.0359	0.2363	0.0295	0.0236
pi,e		1.4870	0.1487	1.3189	0.1319	1.4093	0.1566	0.1409	1.6071	0.2009	0.1607
$\hat{\sigma}^2_{Rel}$		1.4870	0.1487	1.3189	0.1319	1.4093	0.1566	0.1409	1.6071	0.2009	0.1607
$\hat{\sigma}^2_{Abs}$		1.7538	0.1754	1.4417	0.1442	1.7678	0.1964	0.1768	1.8434	0.2304	0.1843
$\hat{\rho}^2$		0.28	0.80	0.32	0.82	0.36	0.84	0.85	0.38	0.83	0.86
$\hat{\phi}$		0.25	0.77	0.30	0.81	0.31	0.80	0.82	0.35	0.81	0.84

Note: n'_i denotes the number of items in each subscale.

generalizability of 0.80, only one rater is needed for mathematics but three raters are needed for reading.

Two-Facet, Partially Nested $p \times (i{:}s)$ Design with s Fixed

In the study of the Self-Description Questionnaire, 140 persons responded to three scales: general self-concept, academic self-concept, and mathematics self-concept. The items differ from scale to scale, so items are nested within scales. Since all persons completed all items in all scales, the design is partially nested, with items (i) nested within scales (s) and both crossed with persons (p). We denote this design as $p \times (i{:}s)$. The scales observed in the G study (general, academic, and mathematics self-concept) were selected purposefully with no intent to generalize beyond them. Hence, the scale facet is properly treated as being fixed. Table 7.6 contains estimated variance components, error variances, and generalizability coefficients for the average of the three scales and for each scale separately. Using estimated error-variance components from the G study, we compute the error-variance components for alternative D studies by dividing by n_i', the number of items in a scale.

If the decision maker intends to use an average score across scales, using 10 items per scale yields minimally acceptable levels of estimated generalizability (Table 7.6). If the decision maker intends to use separate scores for each scale, using 10 items in each scale would produce scale scores with fairly good estimated generalizability. If the user wants to shorten the questionnaire and is satisfied with, say, a coefficient for absolute decisions of 0.80, the academic and mathematics scales can be shortened to 9 items and 8 items, respectively.

Summary

Generalizability (G) studies provide estimates of components of universe-score variance and error variance. Decision (D) studies use G study variance-component information to design and evaluate measurement procedures for practical applications. Alternative measurement procedures are identified by varying the number of conditions of each facet. These alternatives are evaluated as to the magnitude of measurement error and generalizability (reliability) that can be expected from each. Just as the Spearman–Brown prophecy formula in classical test theory permits the decision maker to tailor the length of a test to meet psychometric standards within the bounds of time and cost, D studies can project, simultaneously, the effect of alternative combinations of items and occasions and other facets needed to meet psychometric standards within time and cost limits.

The general strategy of design and evaluation can be exemplified with a simple one-facet, persons-by-items crossed design. Alternative measurement procedures—in this case test lengths—can be identified by changing the number of items on the test. To evaluate the impact of test length on relative decisions, measurement error can be estimated for each test length: $\hat{\sigma}^2_{pi,e} / n'_i$. The impact of different numbers of items (n'_i) can be evaluated with respect to the magnitude of measurement error ($\hat{\sigma}^2_{pi,e} / n'_i$), and reliability (generalizability) $\hat{\sigma}^2_p$ / $[\hat{\sigma}^2_p + (\hat{\sigma}^2_{pi,e} / n'_i)]$. By adding items to a test, measurement error is reduced and reliability is increased.

But this method is not new. So far, the D study is nothing more than a name for application of the Spearman–Brown formula. G theory, however, moves beyond the capability of classical theory in that it can evaluate the impact of absolute interpretations of measurements, and the trade-off between adding conditions of one facet versus another in multifaceted measurements. With respect to absolute decisions, the same procedures used for relative decisions can be applied. For the one-facet, $p \times i$ design, measurement error is defined as $\hat{\sigma}^2_i / n'_i + \hat{\sigma}^2_{pi,e} / n'_i$. The impact of different test lengths (numbers of items) on measurement error and dependability $\hat{\sigma}^2_p / [\hat{\sigma}^2_p + (\hat{\sigma}^2_i / n'_i) + (\hat{\sigma}^2_{pi,e} / n'_i)]$ can be evaluated not only psychometrically but also in cost and time terms.

With respect to multifaceted designs, D studies evaluate the impact of increasing the number of conditions of two or more facets simultaneously. For example, in a persons-by-items-by-occasions design, the D study explores the trade-off between adding items and increasing the number of occasions on which the measurement is taken. For relative decisions the magnitude of measurement error $[(\hat{\sigma}^2_{pi} / n'_i) + (\hat{\sigma}^2_{po} / n'_o) + (\hat{\sigma}^2_{pio,e} / n'_i n'_o)]$ and generalizability $\{ \hat{\sigma}^2_p / [\hat{\sigma}^2_p + (\hat{\sigma}^2_{pi} / n'_i) + (\hat{\sigma}^2_{po} / n'_o) + (\hat{\sigma}^2_{pio,e} / n'_i n'_o)] \}$ can be determined for alternative measurement designs that vary the number of items and occasions. In a similar way, alternative two-facet measurement designs for absolute decisions can be evaluated. A particular measurement design that meets psychometric and feasibility criteria then can be selected.

In an analogous manner, D studies can evaluate measurements with nested and fixed facets. The strategy is always the same: project alternative measurements for practical application, and evaluate them in psychometric and feasibility (time, cost) terms.

Exercises

1. In a G study of job performance of machinist mates in the Navy, sailors were evaluated by two observers as they carried out 11 tasks concerning knowledge of gauges, equipment operation, and casualty control aboard ship (Webb, Shavelson, Kim, & Chen, 1989). The estimated variance components for this fully random persons- (p) by-observers- (o) by-tasks- (t) design were $\hat{\sigma}_p^2 = 0.00626$, $\hat{\sigma}_o^2 = 0.00000$, $\hat{\sigma}_t^2 = 0.00970$, $\hat{\sigma}_{po}^2 = 0.00000$, $\hat{\sigma}_{pt}^2 = 0.02584$, $\hat{\sigma}_{ot}^2 = 0.00003$, $\hat{\sigma}_{pot,e}^2 = 0.00146$.
   a. What do the results of the G study say about the sources of variability and implications for the design of the D study?
   b. Calculate the estimated variance components and error variance and generalizability coefficients for relative and absolute decisions for $n'_o = 1, n'_t = 1$; $n'_o = 1, n'_t = 11$; $n'_o = 2, n'_t = 11$; and $n'_o = 1, n'_t = 17$. Compare the results.
2. For the study of Marine Corps infantry riflemen evaluated while assembling communications equipment (chapter 4, Exercises 2 and 3), calculate the estimated variance components and generalizability coefficients for n'_r and n'_o pairs (1, 1), (2, 3), (3, 2), (2, 4), and (4 2). Compare the results.
3. The Beginning Teacher Evaluation Study (BTES), an 8-year research program to identify effective teaching behavior in elementary school reading and mathematics, administered test batteries to second- and fifth-grade students on multiple occasions. In a G study three arithmetic subtests (addition/subtraction, multiplication, and division: 10 items each) were administered to 127 fifth-grade students on two occasions. Tables 7.7 and 7.8 show the results for the design treated as fully random and for the subtest facet treated as fixed (the average over the three subtests, each subtest separately; Webb, Shavelson, & Maddahian, 1983).
   a. Interpret the magnitude of variability of the fixed facet subtest.
   b. Calculate the estimated variance components, error variances, and G coefficients for the average of the three subtests and for each subtest separately for 10 items and two occasions.
   c. Interpret and compare the results in (b).

TABLE 7.7 Estimated Variance Components for Fully Random $p \times o \times (i{:}s)$ Design

Source of Variation	$n'_s = 1, n'_i = 1, n'_o = 1$
Persons (p)	0.018
Occasions (o)	0.001
Subtests (s)	0.021
Items:Subtests ($i{:}s$)	0.008
po	0.004
ps	0.018
os	0.004
pi:s	0.018
oi:s	0.001
pos	0.019
poi:s,e	0.101

TABLE 7.8 Estimated Variance Components for $p \times o \times i$ Design ($n'_o = 1, n'_i = 1$)

Source of Variation	*Average over 3 Subtests*	*Addition/ Subtraction*	*Multiplication*	*Division*
Persons (p)	0.024	0.022	0.054	0.034
Occasions (o)	0.002	0.000	0.012	0.002
Items (i)	0.003	0.010	0.007	0.008
po	0.010	0.012	0.047	0.009
pi	0.006	0.010	0.022	0.024
oi	0.000	0.000	0.002	0.001
poi,e	0.034	0.109	0.112	0.081

Answers to Exercises

1. a. Observer was a negligible source of variation: Most variance components associated with observer are zero or near zero. The task, on the other hand, was a major source of variation. Some tasks were easier than others (large t effect), and sailors' relative standing differed across tasks (large $p \times t$ interaction effect). The D study need only use a single observer (and it does not matter whether the observer is the same for all persons and tasks), but it should use multiple tasks.

   b. See Table 7.9. A performance test consisting of a single task has poor generalizability; even a test consisting of 17 tasks would not produce very high estimated generalizability. Using two observers yields no advantage over using one.

TABLE 7.9

Source	n'_o = / n'_t =	1 / 1	1 / 11	2 / 11	1 / 17
Persons (p)		0.00626	0.00626	0.00626	0.00626
Observers (o)		0.00000	0.00000	0.00000	0.00000
Tasks (t)		0.00970	0.00088	0.00088	0.00057
po		0.00000	0.00000	0.00000	0.00000
pt		0.02584	0.00235	0.00235	0.00152
ot		0.00003	0.00000	0.00000	0.00000
pot,e		0.00146	0.00013	0.00007	0.00009
$\hat{\sigma}^2_{\text{Rel}}$		0.02730	0.00248	0.00242	0.00161
$\hat{\sigma}^2_{\text{Abs}}$		0.03703	0.00337	0.00330	0.00218
$\hat{\rho}^2$		0.19	0.72	0.72	0.80
$\hat{\phi}$		0.14	0.65	0.65	0.74

2. See Table 7.10. For the same number of observations per rifleman, slightly higher generalizability is obtained with more occasions and fewer raters. It is also important, however, to balance practical considerations in the choice of design. If it is less expensive to evaluate riflemen on fewer occasions but with more raters, the decision maker may choose to use such a design.
3. a. The large subtest main effect (0.021) shows that some subtests were more difficult than others. The large person-by-subtest interaction (0.018) shows that persons' relative standing differed across subtests. The large person-by-item interaction (0.018) shows that persons' relative standing differed across items. The large person-by-occasion-by-subtest interaction shows that the relative standing of persons across occasions differed from one subtest to another. Together, these effects suggest that student achievement differed considerably across subtests. Analyses of each subtest separately may illuminate different patterns of student performance.

   b. See Table 7.11.

   c. (a) More variability occurred across occasions for multiplication than the other subtests, and least error overall for the division subtest. (b) Lengthening the addition/subtraction and multiplication subtests and/or administering them on more occasions would increase their levels of generalizability. (c) The average over the three subtests yields estimated generalizability coefficients nearly as large as those for division.

TABLE 7.10

Source of Variation	n'_r = / n'_o =	*1* / *1*	*2* / *3*	*3* / *2*	*2* / *4*	*4* / *2*
Persons (p)		0.0125	0.0125	0.0125	0.0125	0.0125
Occasions (o)		0.0070	0.0023	0.0035	0.0018	0.0035
Raters:Occasions ($r{:}o$)		0.0013	0.0002	0.0002	0.0002	0.0002
po		0.0035	0.0012	0.0018	0.0009	0.0018
pr,pro,e		0.0154	0.0026	0.0026	0.0019	0.0019
$\hat{\sigma}^2_{\text{Rel}}$		0.0189	0.0038	0.0044	0.0028	0.0037
$\hat{\sigma}^2_{\text{Abs}}$		0.0272	0.0063	0.0081	0.0048	0.0074
$\hat{\rho}^2$		0.40	0.77	0.74	0.82	0.77
$\hat{\phi}$		0.31	0.66	0.61	0.72	0.63

TABLE 7.11 Estimated Coefficients for $n'_o = 2, n'_i = 10$

Source of Variation	*Average over 3 Subtests*	*Addition/ Subtraction*	*Multiplication*	*Division*
Persons (p)	0.024	0.022	0.054	0.034
Occasions (o)	0.001	0.000	0.006	0.001
Items (i)	0.000	0.001	0.001	0.001
po	0.005	0.006	0.024	0.005
pi	0.001	0.001	0.002	0.002
oi	0.000	0.000	0.000	0.000
poi,e	0.002	0.005	0.006	0.004
$\hat{\sigma}^2_{\text{Rel}}$	0.008	0.012	0.032	0.010
$\hat{\sigma}^2_{\text{Abs}}$	0.009	0.013	0.039	0.012
$\hat{\rho}^2$	0.75	0.65	0.63	0.77
$\hat{\phi}$	0.73	0.63	0.58	0.74

Note

1. We do not divide $\hat{\sigma}^2_{pi,e}$ by n_p because persons is the object of measurement. Attention focuses on individuals' scores, not on group means (see chapter 6).

8

Generalizability and Decision Studies with Different Designs

The results of a crossed, random-effects G study can be used to evaluate a large number of alternative D-study designs with crossed or nested, random or fixed, facets. More specifically, variance-component estimates from a crossed G study can be used to estimate the variance components, error variances, and generalizability coefficients for D studies with different designs from that of the G study. The most flexible G studies will have crossed designs with random facets. A nested G study cannot be used to estimate effects in a crossed D study due to confounding. Moreover, it is not sensible to use a G study with a fixed facet to estimate effects for a D study with that facet random. G studies with nested and/or fixed facets cannot be applied to as many D-study designs as crossed, random G studies.

Even when the G study has a crossed, random design, for a variety of reasons it may be advisable to use a nested D study or one with a fixed facet. Not only does a nested D-study measurement sometimes have higher estimated generalizability than a crossed design with the same number of observations (as will be discussed below), but it may be easier or less costly to implement. For example, in a study of teacher behavior it may be easier to have different raters observe different teachers than to have the same raters observe all teachers. The same principle would apply to any observational or interview situation in which having a set of raters or interviewers observe all persons may be logistically difficult or too expensive. The decision maker also may fix a facet in a D study that was random in the G study. For example, he or she may decide to restrict the set of observers in the D study to those used in the G study.

This chapter contains general principles and examples of using the results of G studies to design a wide variety of alternative D studies. Specifically, it shows how to estimate variance components, error variances, and generalizability coefficients for D studies with designs different from that used in the G study.

One-Facet, Crossed G-Study and Nested D-Study (Random) Designs

For ease, a one-facet design is a good starting point to show how variance components from a crossed design can be combined to form the variance components in a nested design. The general principle for combining variance components is the following: The variance component for confounded effects in a nested design is the sum of the variance components for the separate effects in the corresponding crossed design.

The variance components in a one-facet, crossed design are σ_p^2, σ_i^2, and $\sigma_{pi,e}^2$. The variance components in a one-facet, nested *i*:*p* design are σ_p^2 and $\sigma_{i,pi,e}^2$. Hence, the residual component in the nested design is the sum of the separate components in the crossed design (σ_i^2 and $\sigma_{pi,e}^2$):

Nested *i*:*p* Design		Crossed $p \times i$ Design	
$\sigma_{i,pi,e}^2$	$=$	$\sigma_i^2 + \sigma_{pi,e}^2$	[8.1]

The variance components that are not confounded with other effects (σ_p^2 in the nested *i*:*p* design) remain the same in crossed and nested designs.

Table 8.1 provides the estimated variance components, error variances, and generalizability coefficients for a crossed $p \times i$ G study and a nested *i*:*p* D study using the science achievement test scores discussed in an earlier example. Because the item effect is quite small ($\hat{\sigma}_i^2 =$ 0.0093), the results for the crossed and nested designs are very similar in this example. The results for absolute decisions are identical in the one-facet nested and crossed designs because absolute error variance is the same in the two designs: $\hat{\sigma}_i^2 / n_i' + \hat{\sigma}_{pi,e}^2 / n_i'$ in the crossed design and $\hat{\sigma}_{i,pi,e}^2 / n_i'$ in the nested design. And the results for relative decisions are very similar because the item effect is quite small in this data set ($\hat{\sigma}_i^2 = 0.0093$). Had the item effect been larger, relative-error variance would have been larger in the nested design than in the crossed design

TABLE 8.1 Comparison of One-Facet, Crossed ($p \times i$) G-Study and One-Facet, Nested ($i{:}p$) D-Study Designs (Random Effects)

Source of Variation	$\hat{\sigma}^2$	Variance Component $n_i' = 1$	20	*Source of Variation*	$\hat{\sigma}^2$	Variance Component $n_i' = 1$	20
Crossed $p \times i$ Design				Nested $i{:}p$ Design			
Persons (p)	$\hat{\sigma}^2_p$	0.0305	0.0305	Persons (p)	$\hat{\sigma}^2_p$	0.0305	0.0305
Items (i)	$\hat{\sigma}^2_i$	0.0093	0.0005	Items:			
pi,e	$\hat{\sigma}^2_{pi,e}$	0.2103	0.0105	Persons ($i{:}p$)	$\hat{\sigma}^2_{i,pi,e}$	0.2196	0.0110
$\hat{\sigma}^2_{\text{Rel}}$		0.2103	0.0105			0.2196	0.0110
$\hat{\sigma}^2_{\text{Abs}}$		0.2196	0.0110			0.2196	0.0110
$\hat{\rho}^2$		0.13	0.74			0.12	0.74
$\hat{\phi}$		0.12	0.74			0.12	0.74

and, hence, the estimated generalizability coefficient would have been smaller. Based on the similarity of the results for the two designs in Table 8.1, if the decision maker wants to administer different sets of items to different people instead of the same items to all people, the level of generalizability will not suffer.

Two-Facet, Crossed G-Study and Nested D-Study Designs

A two-facet, crossed G study can be used to estimate variance components, error variances, and generalizability coefficients for a wide variety of nested decision-study designs: $p \times i{:}j$ or $p \times j{:}i$, $i{:}p \times j$ or $j{:}p \times i$, $i{:}(p \times j)$ or $j{:}(p \times i)$, $(i \times j){:}p$, and $i{:}j{:}p$ or $j{:}i{:}p$. The general principles for estimating variance components in nested or partially nested D studies from a crossed G study are the same as outlined in the previous section: (a) The variance component for each confounded effect in the D study is the sum of the separate variance components from the crossed G study, and (b) the variance components for the nonconfounded effects remain the same.

Consider, for example, the two-facet crossed G study of students' help-seeking behavior described earlier. This G study has a persons- (p) by-raters- (r) by-occasions- (o) design. The decision maker may wish to use different raters on each occasion but still have each person rated by all raters on each and every occasion. This D study would have a $p \times (r{:}o)$ design. As was shown in chapter 4, this design has five variance components: σ^2_p, σ^2_o, σ^2_{po}, $\sigma^2_{r,ro}$, and $\sigma^2_{pr,pro,e}$. The variance components with confounded effects in this design are $\sigma^2_{r,ro}$ and $\sigma^2_{pr,pro,e}$. They can be determined easily from the variance components in the crossed G study:

$p \times (r{:}o)$ design		$p \times r \times o$ design	
$\sigma^2_{r,ro}$	$=$	$\sigma^2_r + \sigma^2_{ro}$	[8.2]
$\sigma^2_{pr,pro,e}$	$=$	$\sigma^2_{pr} + \sigma^2_{pro,e}$	

The remaining variance components, σ^2_p, σ^2_o, and σ^2_{po} are the same in both designs. Table 8.2 illustrates how the $p \times r \times o$ G study can be used to estimate the generalizability of a $p \times (r{:}o)$ D study. The estimated generalizability is slightly higher for the nested design than for the crossed design. When the number of occasions is greater than 1, the error variances in the nested design are reduced by virtue of the larger denominators. For example, in the crossed design, $\hat{\sigma}^2_r$ is divided by the number of raters in the D study, n'_r. In the partially nested design, in contrast, $\hat{\sigma}^2_r$ is divided by the number of raters *and* the number of occasions in the D study ($n'_r\, n'_o$) by virtue of its being confounded with $\hat{\sigma}^2_{ro}$. Similarly, $\hat{\sigma}^2_{pr}$ is divided by the number of raters and the number of occasions by virtue of its being confounded with $\hat{\sigma}^2_{pro,e}$. In terms of estimated generalizability, then, the partially nested $p \times (r{:}o)$ design has an advantage over the fully crossed $p \times r \times o$ design. The larger the effects due to raters, the greater will be the advantage of the nested design.

Two-Facet, Random G and D Studies: General Principles

Any G study can be used to estimate the effects in a D-study design with the same or more nesting than in the G-study design. Table 8.3 lists the possible two-facet G- and D-study designs. The principle for estimating variance components is the same as described above: (a) for confounded effects in the D study, sum the variance components from

TABLE 8.2 Comparison of Two-Facet, Crossed $p \times r \times o$ and Two-Facet, Partially Nested D-Study Designs (Random Effects)

Source of Variation	$\hat{\sigma}^2$	$n'_r = 1$, $n'_o = 1$	$n'_r = 2$, $n'_o = 4$	Source of Variation	$\hat{\sigma}^2$	$n'_r = 1$, $n'_o = 1$	$n'_r = 2$, $n'_o = 4$
Crossed $p \times o$ Design				Partially Nested $p \times (r{:}o)$ Design			
Persons (p)	$\hat{\sigma}^2_p$	0.3974	0.3974	Persons (p)	$\hat{\sigma}^2_p$	0.3974	0.3974
Occasions (o)	$\hat{\sigma}^2_o$	0.1090	0.0273	Occasions (o)	$\hat{\sigma}^2_o$	0.1090	0.0273
Raters (r)	$\hat{\sigma}^2_r$	0.0096	0.0048	Raters:			
ro	$\hat{\sigma}^2_{ro}$	0.0064	0.0008	Occasions ($r{:}o$)	$\hat{\sigma}^2_{r,ro}$	0.0160	0.0020
po	$\hat{\sigma}^2_{po}$	0.3141	0.0785	*po*	$\hat{\sigma}^2_{po}$	0.3141	0.0785
pr	$\hat{\sigma}^2_{pr}$	0.0673	0.0337	*pr:o,e*	$\hat{\sigma}^2_{pr,pro,e}$	0.2917	0.0365
pro,e	$\hat{\sigma}^2_{pro,e}$	0.2244	0.0281				
$\hat{\sigma}^2_{\text{Rel}}$		0.6058	0.1403			0.6058	0.1150
$\hat{\sigma}^2_{\text{Abs}}$		0.7308	0.1732			0.7308	0.1443
$\hat{\rho}^2$		0.40	0.74			0.40	0.78
$\hat{\phi}$		0.35	0.70			0.35	0.73

the crossed G study; and (b) leave the remaining effects in the D study alone.

Two-Facet, Crossed Design: Random Versus Mixed

The decision maker may decide to fix a facet in the D study that was random in the G study. The appropriate analysis is to average over the conditions of the fixed facet in the analysis of the random portion of the design. Consider, for example, the two-facet, crossed study of children's help-seeking behavior discussed earlier, in which children were rated by two raters on two occasions. The design has persons (p), raters (r), and occasions (o) crossed, which is denoted as $p \times r \times o$. If raters are considered random in the D study, then the decision maker can generalize to the average over all raters in a large universe of raters. The decision maker may instead decide to use the same set of raters in future

TABLE 8.3 Possible Random D-Study Designs from a Random, Two-Facet G-Study Design

G-Study Design	*D-Study Design*		
$p \times i \times j$	$p \times (i{:}j)$	or	$p \times (j{:}i)$
	$(i{:}p) \times j$	or	$(j{:}p) \times i$
	$i{:}(p \times j)$	or	$j{:}(p \times i)$
	$(i \times j){:}p$		
	$i{:}j{:}p$	or	$j{:}i{:}p$
$p \times (i{:}j)$	$i{:}j{:}p$		
$p \times (j{:}i)$	$j{:}i{:}p$		
$i{:}(p \times j)$	$i{:}j{:}p$		
$j{:}(p \times i)$	$j{:}i{:}p$		
$(i \times j){:}p$	$i{:}j{:}p$	or	$j{:}i{:}p$

D studies. In this case, the decision maker does not plan to generalize beyond these raters and so this facet is fixed. In this scenario, averaging over raters makes sense, while analyzing each rater separately (the alternative approach to analyzing designs with a fixed facet, see chapter 5) does not. The decision maker may choose to limit generalization to these two raters, but has no interest in the measurement for each rater separately.

The $p \times r \times o$ design with raters fixed resolves into a one-facet design with persons crossed with occasions. Follow the procedures given in chapter 5 when averaging over conditions of the fixed facet; the variance components are calculated as follows:

$$\hat{\sigma}^2_{p*} = \hat{\sigma}^2_p + \frac{1}{n_r}\,\hat{\sigma}^2_{pr} \tag{8.3}$$

$$\hat{\sigma}^2_{o*} = \hat{\sigma}^2_o + \frac{1}{n_r}\,\hat{\sigma}^2_{ro} \tag{8.4}$$

$$\hat{\sigma}^2_{po,e*} = \hat{\sigma}^2_{po} + \frac{1}{n_r}\,\hat{\sigma}^2_{pro,e} \tag{8.5}$$

The variance components on the left-hand side of Equations 8.3, 8.4, and 8.5 are those in the mixed design; those on the right-hand side are

TABLE 8.4 Comparison of Two-Facet, Crossed $p \times r \times o$ Designs: D Studies with r Random versus r Fixed

Source of Variation	$n'_r =$ $n'_o =$	*1* *1*	2 *1*	2 4	*Source of Variation*	$n'_o =$	*1*	4
Design with r Random					Design with r Fixed[a]			
Persons (p)		0.3974	0.3974	0.3974	Persons (p^*)		0.4311	0.4311
Raters (r)		0.0096	0.0048	0.0048				
Occasions (o)		0.1090	0.1090	0.0273	Occasions (o^*)		0.1122	0.0281
pr		0.0673	0.0337	0.0337				
po		0.3141	0.3141	0.0785	*po,e**		0.4263	0.1066
ro		0.0064	0.0032	0.0008				
pro,e		0.2244	0.1122	0.0281				
$\hat{\sigma}^2_{Rel}$		0.6058	0.4600	0.1403			0.4263	0.1066
$\hat{\sigma}^2_{Abs}$		0.7308	0.5770	0.1732			0.5385	0.1347
$\hat{\rho}^2$		0.40	0.46	0.74			0.50	0.80
$\hat{\phi}$		0.35	0.41	0.70			0.44	0.76

Note: n'_r = number of raters and n'_o = number of occasions in the D study.
a. Averaging over two raters.

from the random-effects G study (first column in Table 8.4); and n_r is the number of raters in the G study, here 2.

Table 8.4 presents the results for the design as fully random and for the design with raters fixed, averaging over the two raters. Because the design with r fixed averages over the two raters, the random design results used for comparison should have two raters ($n'_r = 2$ in Table 8.4). As can be seen in Table 8.4, whether to treat raters as random or fixed has implications for estimated generalizability in this study. Averaging over two raters and using four occasions yields estimated generalizability and phi coefficients of 0.80 and 0.76 when raters are considered fixed, and 0.74 and 0.70 when they are treated as random.

When the rater facet is treated as fixed, variation in children's scores across raters is not considered error variance. The design treating rater as fixed assumes that the score of interest is the average over the two raters. So a child's universe score is the average over the two raters because these raters constitute the entire universe of raters. No error is found in generalizing from the average over these raters to the universe of raters. Error is found only in generalizing from the occasions used here to all occasions in the universe.

The choice between mixed and random designs involves a trade-off. The mixed design produces higher estimated generalizability coefficients, but the scores are generalized only to a restricted universe. Any conclusions about children's behavior pertain only to scores produced by these two raters. The random design produces lower levels of estimated generalizability, but the scores are generalized to a wide universe. When choosing between designs, the decision maker must balance these concerns.

Two-Facet, Nested Design: Random Versus Mixed

Consider the teacher behavior study discussed earlier in which five teachers were observed by three raters on three occasions (Erlich & Shavelson, 1976). The occasions were different for each teacher, but all raters coded the behavior of all teachers on all occasions. In this design, occasions are nested within teachers, and both occasions and teachers are crossed with raters, which is denoted as $(o{:}p) \times r$. In the analyses presented earlier, raters was treated as a random facet. If the decision maker decided instead to use the same set of three raters in future D studies, raters would be considered a fixed facet.

The $(o{:}p) \times r$ design with raters fixed resolves into a one-facet design with occasions nested within persons. From the procedures given in chapter 5, the variance components in the mixed design are calculated as follows:

$$\hat{\sigma}^2_{p*} = \hat{\sigma}^2_{p} + \frac{1}{n_r}\,\hat{\sigma}^2_{pr} \qquad [8.6]$$

$$\hat{\sigma}^2_{o,po*} = \hat{\sigma}^2_{o,po} + \frac{1}{n_r}\,\hat{\sigma}^2_{ro,pro,e} \qquad [8.7]$$

The variance components on the left-hand side of Equations 8.6 and 8.7 are those in the mixed design; the variance components on the right-hand side are from the random-effects G study (first column in Table 8.5); and n_r is the number of raters in the G study, here 3.

TABLE 8.5 Comparison of Two-Facet, Nested $(o{:}p) \times r$ Designs: D Studies with r Random Versus r Fixed

Source of Variation	$n'_r = 1$, $n'_o = 1$	$n'_r = 3$, $n'_o = 1$	$n'_r = 3$, $n'_o = 3$	*Source of Variation*	$n'_o = 1$	$n'_o = 3$
Design with r Random				Design with r Fixed[a]		
Teachers (p)	8.288	8.288	8.288	Teachers (p^*)	8.800	8.800
Raters (r)	2.613	0.871	0.871			
Occasions:				Occasions:		
Teachers ($o{:}p$)	4.622	4.622	1.541	Teachers ($o{:}p$)*	7.646	2.549
pr	1.537	0.512	0.512			
$(o{:}p)r,e$	9.072	3.024	1.008			
$\hat{\sigma}^2_{\text{Rel}}$	15.231	8.158	3.061		7.646	2.549
$\hat{\sigma}^2_{\text{Abs}}$	17.844	9.029	3.932		7.646	2.549
$\hat{\rho}^2$	0.35	0.50	0.73		0.54	0.78
$\hat{\phi}$	0.32	0.48	0.68		0.54	0.78

Note: n'_r = number of raters and n'_o = number of occasions in the D study.
a. Averaging over raters.

Table 8.5 presents the results for the fully random design and for the design with rater fixed (averaging over the three raters). As before, fixing the rater facet produces somewhat higher estimated generalizability coefficients. But also as before, the cost is a more restricted generalization of the scores. Any conclusions about teacher behavior pertain only to scores produced by those three raters.

Summary

The results of crossed G studies can be used to design D studies with crossed or nested, random or fixed, facets. This flexibility provides the decision maker with the capability of tailoring a measurement procedure to his or her purpose, trading off generalizability, time, and cost.

In designing alternative D studies with nested facets from a crossed G study, two principles apply: (a) The variance component for confounded effects in a nested design is the sum of the variance components for the separate effects in the crossed design, and (b) the variance components for nonconfounded effects remain the same as in the crossed G-study design. Examples of one- and two-facet, nested D studies are presented here.

D studies also may contain a fixed facet even though this facet was random in the G study. By fixing a facet, measurement error is diminished because the researcher does not generalize from the conditions of the fixed facet to additional unobserved conditions. The procedures for analyzing models with fixed facets were presented in chapter 5. They are applied to several two-facet D studies in this chapter.

Exercises

1. In a G study of the U.S. Department of Labor's ratings of educational requirements of occupations, 71 field analysts used the General Education Development (GED) scale to evaluate the reasoning, mathematics, and language requirements of 27 occupations. The raters (r) were given written descriptions of the 27 jobs (j) and carried out their ratings on two occasions (o). Because interest lay in assessing systematic differences between jobs with respect to the general educational development required to perform them, the object of measurement was jobs. (G theory has great flexibility in the choice of object measurement. Although the typical application is the differentiation of individuals, any facet may be the focus of study. This wide applicability is called the *principle of symmetry;* Cardinet, Tourneur, & Allal, 1976.) The design of the G study was fully crossed, and all facets were random. The results for the ratings of reasoning are given in Table 8.6.
   a. Interpret the results.
   b. How many occasions and raters are needed for estimated relative and absolute G coefficients of at least 0.85?
   c. In the decision study, the educational requirements of thousands of jobs will be evaluated. Describe the practical advantages of a D-study design in which raters are nested within jobs (r:j).
   d. Calculate the variance components for a D study with design $(r{:}j) \times o$.
   e. Compare the results of the $j \times o \times r$ D study with those of the $(r{:}j) \times o$ D study.
2. In the study of job performance of Navy machinist mates (chapter 7, Exercise 1), suppose that task is treated as a fixed facet in the D study. Calculate the estimated variance components and generalizability coefficients averaging over tasks.

TABLE 8.6

Source of Variation	*Sum of Squares*	*df*	*Mean Square*	$\hat{\sigma}^2$
Jobs (*j*)	2817.100	26	108.350	0.760
Occasions (*o*)	0.730	1	0.730	0.000
Raters (*r*)	256.900	70	3.670	0.050
jo	4.940	26	0.190	0.000
jr	782.600	1820	0.430	0.120
or	51.100	70	0.730	0.020
jor,e	345.800	1820	0.190	0.190

Source: The data are from "Generalizability of General Educational Development Ratings of Jobs in the U.S." by N. M. Webb, R. J. Shavelson, J. Shea, and E. Morello, *Journal of Applied Psychology*, 1981, *66*.

Answers to Exercises

1. a. Little variation was due to occasions. Raters were a more substantial source of variation. Although raters disagreed little in their mean ratings over job and occasions (small *r* effect), they disagreed more on the relative standing of jobs in reasoning ability required (nonnegligible $j \times r$ effect). The D study should use several raters, but raters need only carry out their evaluations once.

   b. Three raters and one occasion would give $\hat{\rho}^2 = 0.88$ and $\hat{\phi} = 0.86$.

   c. It would be difficult, if not impossible, for the same set of raters to evaluate thousands of jobs (crossed design). In the nested design, each job would be evaluated by a different set of raters. Because the nested design would require thousands of raters, a more reasonable variation of the nested design is to have different sets of jobs evaluated by different sets of raters. Although this design is not, strictly speaking, nested, the variance components from the nested design are a good approximation of those in this design.

   d. $\hat{\sigma}^2_{r,jr}$ and $\hat{\sigma}^2_{or,jor,e}$ are the sum of the corresponding variance components from the crossed design. $\hat{\sigma}^2_j$, $\hat{\sigma}^2_o$, $\hat{\sigma}^2_{jo}$ are the same as those in the crossed design.

   $$\hat{\sigma}^2_{r,jr} = 0.170; \hat{\sigma}^2_{or,jor,e} = 0.210; \hat{\sigma}^2_j = 0.759; \hat{\sigma}^2_o = 0.000; \hat{\sigma}^2_{jo} = 0.000$$

   e. See Table 8.7. In this study, the two designs yield comparable levels of generalizability.

TABLE 8.7

Source of Variation	$n'_r =$ $n'_o =$	*1* *1*	*3* *1*	*Source of Variation*	$n'_r =$ $n'_o =$	*1* *1*	*3* *1*
$j \times o \times r$ Design				$(r{:}j) \times o$ Design			
Jobs (j)		0.760	0.760	Jobs (j)		0.760	0.760
Occasions (o)		0.000	0.000	Occasions (o)		0.000	0.000
Raters (r)		0.050	0.017	Raters (r,jr)		0.170	0.057
jo		0.000	0.000	jo		0.000	0.000
jr		0.120	0.040				
or		0.020	0.007	or,jor,e		0.210	0.070
jor,e		0.190	0.063				
$\hat{\sigma}^2_{Rel}$		0.310	0.103			0.380	0.127
$\hat{\sigma}^2_{Abs}$		0.380	0.127			0.380	0.127
$\hat{\rho}^2$		0.71	0.88			0.67	0.86
$\hat{\phi}$		0.67	0.86			0.67	0.86

2. See Table 8.8. When task is treated as a fixed facet, the decision maker does not generalize beyond the 11 tasks used here and, consequently, no error of generalization due to tasks occurs. The only remaining source of error is observers. Because observers agreed so highly in this study, the levels of generalizability for the mixed design are very high.

TABLE 8.8

Source	$n'_o =$	*1*	2
Persons (p*)		0.00861	0.00861
Observers (o*)		0.00000	0.00000
po,e*		0.00013	0.00007
$\hat{\sigma}^2_{Rel}$		0.00013	0.00007
$\hat{\sigma}^2_{Abs}$		0.00013	0.00007
$\hat{\rho}^2$		0.99	0.99
$\hat{\phi}$		0.99	0.99

9

Summary and Next Steps

In this chapter we briefly summarize the central ideas in G theory that were developed in the *Primer*. We also point to some advanced topics for those who wish to proceed further with generalizability theory.

Summary

Generalizability (G) theory provides a flexible framework for examining the dependability of behavioral measurements. The theory assumes that a measurement taken on a person is only a random sample of that person's behavior. The usefulness of the measurement depends on the degree to which that sample allows us to generalize accurately to the behavior of the same person in a wider set of situations. The concept of reliability, so fundamental to classical theory, is replaced in G theory by the broader notion of generalizability. Instead of asking how accurately a set of observed scores reflects their corresponding true scores, G theory asks how accurately a set of observed scores permits us to generalize about a person's behavior in a defined universe of situations.

G theory extends classical test theory in four important ways. First, the theory estimates statistically the magnitude of each source of error separately in one analysis and provides a mechanism for optimizing the reliability of the measurement (analogous to the Spearman–Brown adjustment for classical test theory). Second, although G theory provides a reliability coefficient, called a "generalizability (G) coefficient," the theory focuses on variance components that index the magnitude of each source of error affecting the measurement. Third, G theory distinguishes between relative decisions and absolute decisions. Relative decisions concern "how much better" one individual performed than another. Absolute interpretations address decisions about "how well" an individual can perform, regardless of his or her peers' performances.

In the latter case, we are concerned about influences that might raise or lower all scores in a group (e.g., an easy form of a test makes the group look good, a difficult form makes the group look bad), as well as those that lead to differences within the group (e.g., Charlie has special knowledge that enabled him to answer correctly certain very difficult questions). Fourth, G theory distinguishes between generalizability (G) and decision (D) studies. G studies estimate the magnitude of as many potential sources of measurement error as possible. D studies use information from a G study to design a measurement that minimizes error for a particular applied purpose. The variance components estimated in the G study, then, are used to design a time- and cost-efficient measurement procedure for widespread application in a D study.

G theory assumes that a person's observed score consists of a universe score (analogous to classical theory's true score) and one or more sources of error. Variability among scores can arise from many possible sources, including differences among individuals (universe-score variation analogous to true-score variation in classical theory), and from multiple sources of error. For example, variability among item scores for a sample of persons arises from differences in (a) the performance of individuals, (b) the difficulty of items, and (c) the interaction of persons and items confounded with other systematic and unsystematic sources.

G theory defines a variance component (σ^2) for each source of variation in observed scores. In our example the variance component for persons, usually the object of measurement, is (σ^2_p), the variance component for items is (σ^2_i), and the variance component for the residual is ($\sigma^2_{pi,e}$). By partitioning variability in this manner, G theory enables the analyst to pinpoint the major sources of measurement error, to estimate the magnitude of each source, to estimate the total magnitude of error, and to form a G coefficient.

G theory applies to a wide variety of measurement designs: crossed, nested, and partially nested, with facets that are random or fixed. In a crossed design, each person responds to the same set of items. A nested variation of this design would use different items for each person, thereby increasing the sample of items from the universe.

Multifaceted measurements may have crossed *and* nested facets, resulting in a partially nested design. Suppose a measure of self-concept that includes 10 items in each of three subscales (general, academic, and mathematics self-concept) is administered to a sample of persons. Persons and items are crossed (all persons respond to all items), but the items are nested within subscales: A different set of 10 items is found in each of the three subscales.

Facets may be random or fixed. A facet is random when the conditions of a facet have been sampled randomly, or the observed conditions are exchangeable with any others in an indefinitely large universe of conditions. A facet is fixed when the conditions are either small in number (e.g., subtests of a general aptitude battery) or the decision maker is interested only in performance under a certain, small number of the conditions (e.g., days within the first week of school) and all conditions are observed in the G study. For example, in studies of elementary school teaching, the focus is on teachers' behavior in certain subject matters. Several raters might observe a set of elementary teachers teaching, say, math lessons and reading lessons. In this case the rater facet may be considered random. The subject matter facet is, however, properly considered fixed. Math and reading exhaust the universe to which a decision maker might wish to generalize.

G theory can be used also to optimize the design of a multifaceted measurement. G studies provide estimates of components of universe-score and error variance. Decision studies use G-study variance-component information to design and evaluate measurement procedures for practical applications. Just as the Spearman–Brown prophecy formula in classical theory permits a decision maker to tailor the length of a test to meet psychometric standards within the bounds of time and cost, D studies can project, simultaneously, the effect of alternative combinations of items and occasions and other facets needed to meet psychometric standards, and time and cost constraints. Not only can decision makers consider designs with different numbers of conditions of the facets, but they can consider a broad array of designs with crossed, nested, random, and fixed facets.

Next Steps

This *Primer* has presented the rudiments of G theory. It has glossed over some important topics (e.g., estimation issues) and has ignored others completely (e.g., multivariate generalizability of profiles). The next steps beyond this *Primer,* then, are to explore more advanced notions in G theory. Our purpose in this closing section is to whet your appetite for more extensive study of G theory and its broad, flexible application to behavioral measurements. To this end, we briefly describe a number of exciting topics to be taken up beyond this book.

Symmetry

Throughout this *Primer,* for simplicity, persons has been the object of measurement. Yet in many behavioral measurements persons are not the object of measurement, but a source of measurement error.

Cardinet, Tourneur, and Allal (e.g., Cardinet & Tourneur, 1985; Cardinet, Tourneur, & Allal, 1981) recognized that the object of measurement may change, depending on a particular decision maker's purpose. Particularly in educational research, interest may attach to rates of success on a particular test item as an indicator of change. Or classrooms, rather than individual students, might be the focus of measurement even though we have a test score for each student in each class. Cardinet et al. (1981) proposed the principle of symmetry to enable G theory to address these situations. They pointed out that any facet may become the object of measurement, depending on the purposes of data collection. Variance components may be computed from a data set regardless of the measurement design; the treatment of the variance components becomes necessarily asymmetric only when a decision is made about the object of measurement.

Generalizability of Multivariate Profiles

Often, more than one score is collected on an individual. For example, intelligence tests produce a profile of scores that includes verbal, quantitative, and spatial abilities. Or in the example of job ratings, each job may be characterized by several attributes, including the reasoning, mathematical, and language development it requires (Webb et al., 1981). In these cases, interest attaches to estimating not only the variance components associated with the object of measurement and the facets of the measurement, but also the covariance of these sources of variability among the set of measures (e.g., verbal, quantitative, spatial ability). G theory handles such multivariate profiles, enabling the user to estimate variance and covariance components, and a multivariate G coefficient (e.g., Cronbach et al., 1972; Shavelson & Webb, 1981; Shavelson, Webb, & Rowley, 1989; Webb et al., 1983).

Variance Component Estimation

This *Primer* has presented one procedure for estimating variance components—least squares (ANOVA) estimation. A number of other methods are available for estimating these components. Recently, maximum likelihood methods have been given considerable attention. Restricted maximum likelihood estimation, for example, provides variance-component estimates that cannot be negative. Marcoulides (1987) found that these estimates performed as well as ANOVA estimates when normal data were used and provided more accurate estimates than the ANOVA estimates when data were ill-conditioned.

Conclusion

These advanced notions extend still further the flexibility of G theory. Even without them, however, G theory remains a powerful technique for examining the dependability of behavioral measurements.

References

Brennan, R. L. (1983). *Elements of generalizability theory.* Iowa City, IA: American College Testing Program.

Brennan, R. L., Jarjoura, D., & Deaton, E. L. (1980). *Some issues concerning the estimation and interpretation of variance components in generalizability theory* (ACT Technical Bulletin No. 36). Iowa City, IA: American College Testing Program.

Brennan, R. L., & Kane, M. T. (1977). An index of dependability for mastery tests. *Journal of Educational Measurement, 14,* 277-289.

Cardinet, J., & Tourneur, Y. (1985). *Assurer la mesure.* Bern, Switzerland: Lang.

Cardinet, J., Tourneur, Y., & Allal, L. (1976). The symmetry of generalizability theory: Application to educational measurement. *Journal of Educational Measurement, 13,* 119-135.

Cardinet, J., Tourneur, Y., & Allal, L. (1981). Extension of generalizability theory and its applications in educational measurement. *Journal of Educational Measurement, 18,* 183-204.

Cornfield, J., & Tukey, J. W. (1956). Average values of mean squares in factorials. *Annals of Mathematical Statistics, 27,* 907-949.

Crick, G. E., & Brennan, R. L. (1982). *GENOVA: A generalized analysis of variance system* (Fortran IV computer program and manual.) Dorchester, MA: University of Massachusetts at Boston, Computer Facilities.

Cronbach, L. J., Gleser, G. C., Nanda, H., & Rajaratnam, N. (1972). *The dependability of behavioral measurements: Theory of generalizability of scores and profiles.* New York: John Wiley. (Available from Books on Demand, University Microfilms, 300 N. Zeeb Rd., Ann Arbor, MI 48106)

Erlich, O., & Borich, C. (1979). Occurrence and generalizability of scores on a classroom interaction instrument. *Journal of Educational Measurement, 16,* 11-18.

Erlich, O., & Shavelson, R. J. (1976). *Application of generalizability theory to the study of teaching* (Technical Report No. 76-9-1). Beginning Teacher Evaluation Study, Far West Laboratory, San Francisco.

Erlich, O., & Shavelson, R. J. (1978). The search for correlations between measures of teacher behavior and student achievement: Measurement problem, conceptualization problem, or both? *Journal of Educational Measurement, 15,* 77-89.

Kane, M. T., & Brennan, R. L. (1977). The generalizability of class means. *Review of Educational Research, 47,* 267-292.

Kenderski, C. M. (1983). *Interaction processes and learning among third-grade black and Mexican-American students in cooperative small groups.* Unpublished doctoral dissertation, University of California, Los Angeles.

Kirk, R. E. (1982). *Experimental design* (2nd ed.). Belmont, CA: Brooks/Cole.

Lord, F. M., & Novick, M. R. (1968). *Statistical theories of mental test scores.* Reading, MA: Addison-Wesley.

Marcoulides, G. A. (1987). *An alternative method for variance component estimation: Applications to generalizability theory.* Unpublished doctoral dissertation, University of California, Los Angeles.

Marsh, H. W. (in press). *Self-description questionnaire.* San Antonio, TX: Psychological Corporation.

Marzano, W. A. (1973). *Determining the reliability of the Distar instructional system observation instrument.* Unpublished master's thesis, University of Illinois, Champaign-Urbana.

Millman, J., & Glass, G. C. (1967). Rules of thumb for writing the ANOVA table. *Journal of Educational Measurement, 4,* 41-51.

Searle, S. R. (1971). *Linear models.* New York: John Wiley.

Shavelson, R. J., Hubner, J. J., & Stanton, G. C. (1976). Self-concept: Validation of construct interpretations. *Review of Educational Research, 46,* 407-441.

Shavelson, R. J., Mayberry, P. W., Li, W., & Webb, N. M. (1990). Generalizability of job performance measurements: Marine Corps rifleman. *Military Psychology, 2,* 129-144.

Shavelson, R. J., Pine, J., Goldman, S. R., Baxter, G. P., & Hine, M. S. (1989, June). New technologies for assessing science achievement. Paper presented at the annual meeting of the American Psychological Society, Washington, DC.

Shavelson, R. J., & Webb, N. M. (1981). Generalizability theory—1973-1980. *British Journal of Mathematical and Statistical Psychology, 34,* 133-166.

Shavelson, R. J., Webb, N. M., & Rowley, G. L. (1989). Generalizability theory. *American Psychologist, 44,* 922-932.

Webb, N. M., Rowley, G. L., & Shavelson, R. J. (1988). Using generalizability theory in counseling and development. *Measurement and Evaluation in Counseling and Development, 21,* 81-90.

Webb, N. M., Shavelson, R. J., Kim, K. S., & Chen, Z. (1989). Reliability (generalizability) of job performance measurements: Navy machinist mates. *Military Psychology, 1,* 91-110.

Webb, N. M., Shavelson, R. J., & Maddahian, E. (1983). Multivariate generalizability theory. *New Directions in Testing and Measurement: Generalizability Theory, 18,* 67-82.

Webb, N. M., Shavelson, R. J., Shea, J., & Morello, E. (1981). Generalizability of general educational development ratings of jobs in the U.S. *Journal of Applied Psychology, 66,* 186-192.

Index

About the Authors

Richard J. Shavelson is Dean of the Graduate School of Education and Professor of Research Methods in the Educational Psychology Program at the University of California, Santa Barbara. He has been Professor of Research Methods at UCLA, Director of the Education and Human Resources Program at the RAND Corporation, and past president of the American Educational Research Association. He is interested in various aspects of measurement theory. His current research focuses on the measurement of individual and group performance in education and military jobs. He is currently exploring alternatives to multiple-choice achievement tests in the assessment of mathematics and science understanding. He has served as principal investigator on three National Science Foundation grants over the past 8 years, all dealing with indicators of the quality of mathematics and science education. He received his Ph.D. in educational psychology from Stanford University.

Noreen M. Webb is Professor of Research Methods and Educational Psychology in the Graduate School of Education at UCLA. She is interested in measurement theory and applications, and her current research focuses on the measurement of learning processes and performance of individuals and groups in educational settings. She is investigating currently the impact of classroom instruction on mathematics problem-solving processes and performance, and is comparing different kinds of tests for measuring mathematics understanding. She recently received a grant from the National Science Foundation to study mathematics learning in different instructional programs and is conducting research on the measurement of mathematics understanding with the Center for Research on Evaluation, Standards, and Student Testing at the UCLA Graduate School of Education. She received her Ph.D. in educational psychology from Stanford University.